MODELING POWER ELECTRONICS AND INTERFACING ENERGY CONVERSION SYSTEMS

MODELING POWER ELECTRONICS AND INTERFACING ENERGY CONVERSION SYSTEMS

M. GODOY SIMÕES
FELIX A. FARRET

WILEY

Published by John Wiley & Sons, Inc., Hoboken, New Jersey
Published simultaneously in Canada

For general information on our other products and services or for technical support, please contact our
Customer Care Department within the United States at (800) 762-2974, outside the United States at
(317) 572-3993 or fax (317) 572-4002.

Wiley also publishes its books in a variety of electronic formats. Some content that appears in print may
not be available in electronic formats. For more information about Wiley products, visit our web site at
www.wiley.com.

Library of Congress Cataloging-in-Publication Data:

Names: Simões, M. Godoy, editor. | Farret, Felix A., editor.
Title: Modeling power electronics and interfacing energy conversion systems /
    edited by Marcelo G. Simões, Felix A. Farret.
Description: Hoboken, New Jersey : John Wiley & Sons, 2017. | Includes bibliographical
    references and index.
Identifiers: LCCN 2016026094 | ISBN 9781119058267 (cloth) | ISBN 9781119058472 (epub)
Subjects: LCSH: Power electronics–Mathematical models. | Energy conversion–Mathematical models.
Classification: LCC TK7881.15 .M58 2016 | DDC 621.31/7–dc23
LC record available at https://lccn.loc.gov/2016026094

Set in 10/12pt Times by SPi Global, Pondicherry, India

Printed in the United States of America

10 9 8 7 6 5 4 3 2 1

CONTENTS

Foreword   **xi**

Preface   **xiii**

1   Introduction to Electrical Engineering Simulation   **1**

    1.1   Fundamentals of State-Space-Based Modeling   4
    1.2   Example of Modeling an Electrical Network   6
    1.3   Transfer Function   9
        1.3.1   State Space to Transfer Function Conversion   10
    1.4   Modeling and Simulation of Energy Systems and Power Electronics   12
    1.5   Suggested Problems   18
    Further Reading   25

2   Analysis of Electrical Circuits with Mesh and Nodal Analysis   **27**

    2.1   Introduction   27
    2.2   Solution of Matrix Equations   28
    2.3   Laboratory Project: Mesh and Nodal Analysis of Electrical Circuits
        with Superposition Theorem   29
    2.4   Suggested Problems   37
    References   40
    Further Reading   40

3   Modeling and Analysis of Electrical Circuits with Block Diagrams   **43**

    3.1   Introduction   43
    3.2   Laboratory Project: Transient Response Study and Laplace
        Transform-Based Analysis Block Diagram Simulation   45

3.3 Comparison with Phasor-Based Steady-State Analysis   52
3.4 Finding the Equivalent Thèvenin   54
3.5 Suggested Problems   56
Further Reading   58

4 Power Electronics: Electrical Circuit-Oriented Simulation   61

4.1 Introduction   61
4.2 Case Study: Half-Wave Rectifier   67
4.3 Laboratory Project: Electrical Circuit Simulation Using PSIM
     and Simscape Power Systems MATLAB Analysis   72
4.4 Suggested Problems   79
Further Reading   81

5 Designing Power Electronic Control Systems   83

5.1 Introduction   83
     5.1.1 Control System Design   85
     5.1.2 Proportional–Integral Closed-Loop Control   86
5.2 Laboratory Project: Design of a DC/DC Boost Converter Control 89
     5.2.1 Ideal Boost Converter   89
     5.2.2 Small Signal Model and Deriving the Transfer Function
            of Boost Converter   90
     5.2.3 Control Block Diagram and Transfer Function   93
5.3 Design of a Type III Compensated Error Amplifier   95
     5.3.1 K Method   95
     5.3.2 Poles and Zeros Placement in the Type III Amplifier   96
5.4 Controller Design   97
5.5 PSIM Simulation Studies for the DC/DC Boost Converter   99
5.6 Boost Converter: Average Model   99
5.7 Full Circuit for the DC/DC Boost Converter   103
5.8 Laboratory Project: Design of a Discrete Control in MATLAB
     Corunning with a DC Motor Model in Simulink   107
5.9 Suggested Problems   112
References   116
Further Reading   116

**6 Instrumentation and Control Interfaces for Energy
    Systems and Power Electronics   117**

6.1 Introduction   117
     6.1.1 Sensors and Transducers for Power Systems Data Acquisition 118
6.2 Passive Electrical Sensors   119
     6.2.1 Resistive Sensors   119
     6.2.2 Capacitive Sensors   121
     6.2.3 Inductive Sensors   123

6.3 Electronic Interface for Computational Data in Power Systems
      and Instrumentation   125
      6.3.1   Operational Amplifiers   125
6.4 Analog Amplifiers for Data Acquisition and Power System Driving   125
      6.4.1   Level Detector or Comparator   126
      6.4.2   Standard Differential Amplifier for Instrumentation and Control   127
      6.4.3   Optically Isolated Amplifier   128
      6.4.4   The V–I Converter of a Single Input and Floating Load   130
      6.4.5   Schmitt Trigger Comparator   131
      6.4.6   Voltage-Controlled Oscillator (VCO)   131
      6.4.7   Phase Shifting   131
      6.4.8   Precision Diode, Precision Rectifier, and the Absolute Value
              Amplifier   134
      6.4.9   High-Gain Amplifier with Low-Value Resistors   136
      6.4.10 Class B Feedback Push–Pull Amplifiers   137
      6.4.11 Triangular Waveform Generator   137
      6.4.12 Sinusoidal Pulse Width Modulation (PWM)   138
6.5 Laboratory Project: Design a PWM Controller with Error Amplifier   140
6.6 Suggested Problems   140
References   145

7 **Modeling Electrical Machines**   **147**

7.1 Introduction to Modeling Electrical Machines   147
7.2 Equivalent Circuit of a Linear Induction Machine Connected
      to the Network   148
7.3 PSIM Block of a Linear IM Connected to the Distribution Network   150
7.4 PSIM Saturated IM Model Connected to the Distribution Network   152
7.5 Doubly Fed Induction Machine Connected to the Distribution
      Network   154
7.6 DC Motor Powering the Shaft of a Self-Excited Induction Generator   156
7.7 Modeling a Permanent Magnet Synchronous Machine (PMSM)   158
7.8 Modeling a Saturated Transformer   158
7.9 Laboratory Project: Transient Response of a Single-Phase Nonideal
      Transformer for Three Types of Power Supply—Sinusoidal, Square
      Wave, and SPWM   158
7.10 Suggested Problems   169
References   175
Further Reading   175

8 **Stand-Alone and Grid-Connected Inverters**   **177**

8.1 Introduction   177
8.2 Constant Current Control   181
8.3 Constant P–Q Control   182
8.4 Constant P–V Control   183

8.5   IEEE 1547 and Associated Controls   184
8.6   P+Resonant Stationary Frame Control   187
8.7   Phase-Locked Loop (PLL) for Grid Synchronization   188
8.8   Laboratory Project: Simulation of a Grid-Connected/
       Stand-Alone Inverter   190
8.9   Suggested Problems   197
References   199
Further Reading   201

9   **Modeling Alternative Sources of Energy**   **203**

9.1   Electrical Modeling of Alternative Power Plants   203
9.2   Modeling a Photovoltaic Power Plant   204
9.3   Modeling an Induction Generator (IG)   205
9.4   Modeling a SEIG Wind Power Plant   207
9.5   Modeling a DFIG Wind Power Plant   208
9.6   Modeling a PMSG Wind Power Plant   208
9.7   Modeling a Fuel Cell Stack   211
9.8   Modeling a Lead Acid Battery Bank   215
9.9   Modeling an Integrated Power Plant   219
9.10   Suggested Problems   224
References   225

10 **Power Quality Analysis**   **227**

10.1   Introduction   227
10.2   Fourier Series   231
10.3   Discrete Fourier Transform for Harmonic Evaluation
       of Electrical Signals   237
       10.3.1 Practical Implementation Issues of DFT Using FFT   237
10.4   Electrical Power and Power Factor Computation for Distorted
       Conditions   239
10.5   Laboratory Project: Design of a DFT-Based Electrical Power
       Evaluation Function in MATLAB   242
10.6   Suggested Problems   250
References   253
Further Reading   253

11 **From PSIM Simulation to Hardware Implementation in DSP**   **255**
    Hua Jin

11.1   Introduction   255
11.2   PSIM Overview   255
11.3   From Analog Control to Digital Control   257
11.4   Automatic Code Generation in PSIM   264
       11.4.1 TI F28335 DSP Peripheral Blocks   265

11.4.2 Adding DSP Peripheral Blocks   266
11.4.3 Defining SCI Blocks for Real-Time Monitoring
        and Debugging   271
11.5 PIL Simulation with PSIM   272
11.6 Conclusion   275
References   278
Further Reading   278

12 Digital Processing Techniques applied to Power Electronics   279

Danilo Iglesias Brandão and Fernando Pinhabel Marafão

12.1 Introduction   279
12.2 Basic Digital Processing Techniques   280
      12.2.1 Instantaneous and Discrete Signal Calculations   280
      12.2.2 Derivative and Integral Value Calculation   280
      12.2.3 Moving Average Filter   282
      12.2.4 Laboratory Project: Active Current Calculation   286
12.3 Fundamental Component Identification   287
      12.3.1 IIR Filter   288
      12.3.2 FIR Filter   290
      12.3.3 Laboratory Project: THD Calculation   291
12.4 Fortescue's Sequence Components Identification   293
      12.4.1 Sequence Component Identification Using IIR Filter   296
      12.4.2 Sequence Component Identification Using DCT Filter   297
      12.4.3 Laboratory Project: Calculation of Negative- and
              Zero-Sequence Factors   298
12.5 Natural Reference Frame PLLs   300
      12.5.1 Single-Phase PLL   301
      12.5.2 Three-Phase PLL   302
      12.5.3 Laboratory Project: Single-Phase PLL Implementation   303
      12.5.4 Laboratory Project: Fundamental Wave Detector
              Based on PLL   306
12.6 MPPT Techniques   307
      12.6.1 Perturb and Observe   310
      12.6.2 Incremental Conductance   310
      12.6.3 Beta Technique   312
      12.6.4 Laboratory Project: Implementing the IC Technique   312
12.7 Islanding Detection   314
      12.7.1 Laboratory Project: Passive Islanding Detection
              Based on IEEE Std. 1547   315
12.8 Suggested Problems   317
References   319

Index   321

FOREWORD

This book is an excellent support to a computer-based course for power electronics, power systems, and alternative energy. All of which are extremely important topics nowadays in electrical engineering. Students and practicing engineers with basic knowledge of transient analysis of electric circuits, energy conversion (electric machinery and transformers), and fundamentals of power electronics or power systems can also benefit by studying this book. The chapters follow a progressive complexity. Every chapter has a brief introduction on the background for the particular content in that chapter; some simple problems are solved; a comprehensive laboratory project is discussed with materials and simulation files available for the reader through a Google Group; and suggested problems can be further developed by instructors, which will enhance the understanding of the chapter topics.

The authors seem to have extensive experience in modeling, simulation, and integration of power electronics in very diverse applications, from circuits to power systems; from machines to generators and turbines; and from renewable energy based on wind, photovoltaics (PVs), hydropower, fuel cells, and geothermal to smart-grid applications. Their expertise made possible the innovative presentation of the advanced topics in the book, from the background knowledge of electric circuits, control of DC/DC converters and inverters, energy conversion, and power electronics. The book prepares readers in applying numerical, analytical, and computational methods for multi-domain simulation of energy systems and power electronics engineering problems.

The sequence in the book starts from an introduction to electrical engineering simulation with analysis of electrical circuits, developing modeling of electrical circuits with linear algebra, block diagrams and circuit analysis, and giving hands-on computational experience for the transient response by Laplace transform-based methodologies. Power electronics circuits are modeled based on electrical circuits and on block diagrams, with examples using PSIM, Matlab®, Matlab/Simulink® and Matlab/Power Systems Toolbox, which has been recently renamed to Simscape Power Systems (and it was before called as SimPowerSystems). Practical implementation of control systems will show students and engineers how to use a computer-oriented approach to design a feedback control for DC/DC converters, DC motors, and stand-alone/grid-connected inverters for wind turbines and PV applications.

The book presents an interesting approach on instrumentation and sensor circuits and systems, with examples using PSIM-based simulations. I enjoyed reading the chapter on modeling electrical machines using equivalent circuits with examples of doubly fed induction machines (DFIMs), self-excited induction generators (SEIGs), permanent magnet synchronous machines and a Simulink-based study on transient modeling of single-phase nonideal transformers with distorted sources. I also appreciated their coverage of modeling alternative sources of energy with several examples of typical plants, such as PV, IG, SEIG, doubly fed induction generator (DFIG), permanent magnet synchronous generator (PMSG), fuel cells, lead acid battery storage, and a case study on modeling an integrated power plant with detailed suggested problems alternative sources of energy. The authors have three complex topics, very well organized in stand-alone and grid-connected inverters with their typical control schemes, discussion of IEEE 1547, PI-resonant control, phase-locked-loop (PLL) for synchronization, with a detailed laboratory project with a comprehensive simulation of a grid-connected/stand-alone inverter.

There is a very authoritative discussion in how to convert PSIM-based simulations in TI-DSP-based hardware co-simulation. The authors gave an in-depth presentation of power quality, Fourier Series, and design of power quality-based scripts for evaluation and designing filters for power systems using discrete Fourier transform (DFT) and Matlab with a very interesting chapter on digital processing techniques applied to power electronics, with several DSP techniques, filters, total harmonic distortion (THD) calculation, single- and three-phase PLLs, and maximum power point tracking (MPPT) techniques with a laboratory project in islanding detection based on IEEE 1547.

The book can be used after an introductory course on power electronics, but it might also be used in a one-semester course with intensive lectures plus laboratory. All the problems, projects, and topics can also be implemented in other computational environments. The theory and methodology presented in the chapters can be easily adapted for other simulation software packages, such as Modelica, PLECS, CASPOC, Simplorer, Saber, Mathematica, or Maple.

I believe that the approach in the book is very innovative. There is no other book available in the market that covers such multi-universe of multi-domain analysis for understanding the computational modeling and analysis of the multidisciplinary topics relevant to power electronics. The book is very modern, and it should be adopted by instructors looking to a new way to teach those advanced concepts.

DR. BIMAL K. BOSE
EMERITUS CHAIR PROFESSOR
DEPARTMENT OF EECS
UNIVERSITY OF TENNESSEE, KNOXVILLE

PREFACE

The book started a few years ago when one of us (M. Godoy Simões) was discussing with Prof. B.K. Bose about the need for an integrated companion textbook for analysis and simulation studies in power electronics, power systems, power quality and renewable energy systems, which is typically a very diverse universe. Our approach could contemplate in just one book the most useful techniques in teaching computational and modeling techniques for those topics. Nowadays, students must be trained and have a multidisciplinary understanding to work in advanced power electronics, and learn to integrate power systems with power electronics, electrome-chanical systems with energy conversion, thermal systems, signal processing, control systems, advanced real-time hardware, DSP, signal processing, mechatronics, renew-able energy, and smart-grid applications. Prof. Bose was a great inspiration to us. He strongly motivated us to work in this project.

Both of us decided to cover in this first edition the foundation of topics that are relevant for a computer-based course for students who have some basic knowledge of power systems and/or power electronics, on modeling power electronics, and interfacing energy conversion systems. Both of us have a solid experience in simulation on practical and theoretical power electronics and energy systems. We have intensive laboratory projects, using Matlab®, Simulink®, Power Systems Toolbox, and PSIM, but the problems can be solved in other simulation environments as well. Chapters 11 and 12 were written by our colleagues who are experts on specific subjects, such as hardware-in-the-loop simulation using PSIM, and applica-tions of digital processing techniques.

The book can support a computer-based laboratory for power electronics, power systems, and alternative energy, as well as serve as a self-study material for readers with background in electrical power who wants to understand how to apply mathematical and engineering tools for modeling, simulation, and control design for energy systems and power electronics. The sequence of chapters follows a progressive complexity, serving as a point of departure for other more complex and detailed power electronics and electrical systems projects. Nevertheless, it is possible to change the order or skip material in order to customize a sequence that fits a combination of the fundamental topics (power electronics, power systems, and renewable energy).

The book was written based on problem-based learning strategies, with a few more complex chapters with project-based learning methodologies. Each chapter has a brief introduction on the theoretical background, a description of the problems to be solved, and objectives to be achieved. Block diagrams, electrical circuits, mathematical analysis, or computer code are discussed with very didactical background lines. Computer solutions for the laboratory projects are discussed, and the simulation files are available for readers who register with the Google Group: Power Electronics Interfacing Energy Conversion Systems, the email address is power-electronics-interfacing-energy-conversion-systems@googlegroups.com. After your registration, you will receive information on how to access the simulation files; the Google Group can also be used to communicate among registered readers of the book.

We build the concepts in the book on the background knowledge of electric circuits, control of DC/DC converters and inverters, energy conversion, and power electronics preparing readers in applying the computational methods for multi-domain simulation of energy systems and power electronics engineering problems. The book can be used for a laboratory with lectures on mathematical analysis and theoretical understanding of several relevant electrical energy conversion systems modeling issues plus laboratory experience in simulation implementation through specific software platforms typically used by industries and research institutions, such as Matlab/Simulink, Power Systems Toolbox, and PSIM, but other computational environments could be used, such as PLECS, CASPOC, Simplorer, Mathematica, and MapleSim.

Chapter 1 gives an introduction to electrical engineering simulation, Chapter 2 covers analysis of electrical circuits with mesh and nodal analysis, and Chapter 3 develops modeling and analysis of electrical circuits with block diagrams, with a laboratory project on the transient response study of Laplace transform-based block diagram systems. Introduction to power electronics is covered in Chapter 4, where an electrical circuit simulation is developed using PSIM and Power Systems Toolbox from Matlab, with Matlab analysis. An in-depth coverage of designing power electronics control systems is made in Chapter 5, with discussions of two projects, how to design a DC/DC boost converter and derive their small signal and transfer function with control implementation plus the study of a discrete control system in Matlab and Simulink of a PI-controlled DC motor drive. Chapter 6 covers detailed instrumentation and sensor circuits and systems, with examples of PSIM-based studies; those circuits can also be implemented in electronics-oriented simulators such as NI/MultiSim, Saber, or Matlab/Simscape. Chapter 7 introduces modeling of electrical machines using equivalent circuits, considering satured magnetic core, with examples of DFIG and DFIM, SEIGs, permanent magnet synchronous machines, and a Simulink-based study on transient modeling of a single-phase nonideal transformers with distorted sources. Chapter 8 covers stand-alone and grid-connected inverters typically used for integrating renewable energy sources, with their typical control schemes, discussion of IEEE 1547, PI-resonant control, PLL for synchronization, with a detailed laboratory project with a comprehensive simulation of a grid-connected/ stand-alone inverter. Chapter 9 has an extensive coverage of modeling alternative sources of energy with several examples of typical plants, such as PV, IG, SEIG, DFIG, PMSG, fuel cells, lead acid battery storage, and a case study on modeling an integrated power plant with detailed suggested problems alternative sources of energy. Power quality is a very important topic, important in both power systems and power electronics approaches. Therefore, Chapter 10 goes in the details of how to use Fourier series, DFT, and fast Fourier transform (FFT), and how to use Matlab for electrical power and power factor computation of distorted conditions. Chapter 11 shows how to use PSIM simulation for hardware implementation in DSP, with several details of using DSP peripheral blocks with PSIM, code generation and processor-in-the-loop (PIL) simulation. Chapter 12 has a comprehensive coverage of digital processing techniques applied to power electronics, with several DSP techniques, filters, calculation of negative and zero-sequence components with their THD calculation. There are discussions on single-phase and three-phase PLLs, and MPPT techniques with a laboratory project in islanding detection based on IEEE 1547.

The real motivation to write this book is that we are living in the twenty-first century and we need to understand how distributed energy and integration of renewable energy to the utility grid can be advanced with information technology, Internet of Things, and artificial intelligence, in a worldwide requirement for renewable energy sources for a sustainable future. The book will prepare students, engineers, and interested readers to contribute to our society with more smart-grid-based applications, more automation, and more control of energy conversion systems.

We thank our colleagues: Dr. Hua Jin (Powersim Inc.), who collaborated with invited Chapter 11 as well as the Brazilian professors, Dr. Danilo Iglesias Brandão (UFMG, Brazil) and Dr. Fernando Pinhabel Marafão (UNESP, Brazil), who collaborated with invited Chapter 12. Dr. Tiago Davi Curi Busarello helped in several case studies and simulations. We acknowledge and are very grateful with the strong support given by Farnaz Harirchi, who developed several simulation cases and prepared many drawings and figures for the book.

We could not have completed this project without the amazing, continued support and encouragement of our friends; the love of our family; and, of course, motivation provided by our dear students.

M. GODOY SIMÕES, DENVER, COLORADO, USA
FELIX A. FARRET, SANTA MARIA, BRAZIL

1

INTRODUCTION TO ELECTRICAL ENGINEERING SIMULATION

Theoretical modeling-based analysis is a process where a model is set up based on laws of nature and logic, using mostly mathematics, physics, and engineering—initially with simplified assumptions about their processes and aiming at finding an input/output model. The following basic procedures and formulations are usually used in supporting a theoretical or an experimental model:

1. Balance equations, for stored masses, energies, and impulses
2. Physical–chemical constitutive equations
3. Phenomenological equations of irreversible processes (thermal conductivity, diffusion, chemical reaction)
4. Entropy balance equations, if several irreversible processes are interrelated
5. Connection equations, describing the interconnection of process elements

Using such formulation principles, a system can be understood in terms of their ordinary differential equations, or their algebraic equations, and then a physical device or a computer simulation or an emulation can be devised in order to obey such equations. The physical system is initialized with their proper initial values, and their development over time mimics the differential equations.

Integrators and function generation can accomplish simulation of an ordinary differential equation (ODE). It has been discussed by Ragazzini in 1947 that the continuous functions of several variables could be approximated by a combination of

Modeling Power Electronics and Interfacing Energy Conversion Systems, First Edition.
M. Godoy Simões and Felix A. Farret.

scalar products, scalar functions, and their time derivatives. We have to find first suitable state variables, i.e. variables that account for energy storage. Typically those variables appear differentiated in the ordinary differential equations.

Several computer-based simulations depend on the principles of analog computing, where a differential equation such as Equation 1.1 must be represented in terms of fundamental operations such as integration, addition, multiplication, and function generation. The old analog computer circuitry required scaling of variables, but in a modern computer, floating-point numbers represents the variables and scaling is not required. Higher precision, flexibility for modifications, better stability, reporting facilities, and lower costs are the main advantages of the digital processing. The analog computing may have an advantage for high-speed online data processing, for example, a voltage across a resistor has immediate response. A function such as the one represented by Equation 1.1 requires several interconnections to represent the required calculations.

$$\frac{dx}{dt} = f(t,x) \tag{1.1}$$

Numerical solution techniques and algorithms to solve differential equation are essential and used in digital computers. There are many ways to find approximate numerical solutions to ordinary differential equations such as the one represented by Equation 1.1. The methods are based on replacing the differential equations by a difference equation. Euler's method is based on the approximation of the derivative by a first-order difference, but there are more efficient techniques such as Runge–Kutta and multistep methods. These methods were well known when digital simulators emerged in the 1960s, but several contributions made them better and more stable when solving difference approximations, for example, the automatic step length adjustment was a very important contribution. A more mathematical-oriented model for dynamical systems can be based on differential–algebraic equations (DAEs), that is, a mixture of differential and algebraic equations, such as those represented by Equation 1.2:

$$g(t, x, \dot{x}) = 0 \tag{1.2}$$

It is not always possible to convert such an equation to an ODE because the Jacobian $\frac{\partial g}{\partial x}$ may not be invertible. Numerical methods for DAEs appeared during the 1970s. However, even until today, the algorithms for DAEs are still not so well developed as the ones for ordinary differential equations. Most of the reliable computer simulators and emulators are based on numerical solution of ODES. So, a DAE is mostly a mathematical exploration, and usually the engineering and physics problems are modeled using ODEs.

When a system is formulated based on DAEs, the derivatives are not usually expressed explicitly. In addition, some derivatives of some dependent variables may not appear in the equations. A system of DAEs can be converted to a system of ODEs by differentiating it with respect to the independent variable. The index of a DAE is effectively the number of times you need to differentiate the DAEs to get a system of ODEs. Even though the differentiation is possible, it is not generally used as a computational technique because properties of the original DAEs are often lost in numerical simulations of the differentiated equations.

Suppose a linear system is defined by an algebraic, such as Equation 1.3.

$$AX = B \tag{1.3}$$

If A is a $m \times n$ matrix, a numerical solution has the following possible scenarios:

- $m = n$, it is a square system, and it can have a unique solution, as long as there are no rows or columns linearly dependent of the other ones. This is usually a numerical problem of matrix inversion. There are interesting input/output mapping of large systems, where A is not known, and experimental data will support the definition of A, for example, with gradient descent methods for system identification.
- $m > n$, it is an overdetermined (or over identified) system and at least one solution can be defined. Overidentified systems are common in curve fitting to experimental data, and least square methods for minimizing the sum of the data deviation squares from the model are a suitable approach.
- $m < n$, it is an underdetermined system, and a trivial solution with at most m nonzero components can be defined. Undetermined systems involve more unknowns than equations, so the solution is never unique. There is a particular solution computed by the so-called QR factorization with a column pivoting method. This kind of problem may have additional constraints, and the methodology becomes the so-called linear programming.

In this book, we emphasize the applications of ODEs, particularly in their state-space format, for modeling energy systems and power electronics. We can then study their dynamics and transient solutions, or we can use linear algebraic systems to understand static or steady-state solutions for such systems. The approach adopted in this book best fits a senior undergraduate or a first-year graduate course. Differential equation-based systems are developed and simulated from practical examples that focus typical electrical circuit applications, energy conversion, renewable energy sources, interconnection of distributed generation, power electronics, power systems, and power quality problems. The linearization of systems is discussed based on average modeling and the use of Taylor series expansion. Techniques of Fourier expansion are developed for power quality considerations, including the understanding of the discrete Fourier transform (DFT), fast Fourier transform (FFT), and wavelet techniques. MATLAB® will be used for programming, solving several numerical algorithms, and graph plotting. Simulink® is used for block diagram-oriented modeling. Electrical circuit-oriented modeling is analyzed using the Power Systems Toolbox of MATLAB as well as the PSIM circuit simulator.

The analog computing paradigm requires explicit state models and linkage from input towards output. That is a kind of limitation because blocks must have a unidirectional data flow from inputs to outputs, but such paradigm supports the majority of solutions for engineering systems. A consequence is that it is difficult to build physics-based model libraries in the block diagram languages with bidirectional dataflow or bidirectional energy flow. There are other more advanced paradigms for simulation of multiphysics domain in object-oriented programming using software for differential–algebraic systems aiming at noncausal modeling with mathematical equations. Such object-oriented approach facilitates the reuse of modeling knowledge. However, this book is not focused on such advanced hybrid computer simulations. The intention of this book is to support a computer-based laboratory for power electronics, power systems, distributed generation, and alternative energy, as well as to be a self-study material for readers with background in electrical power who want to understand how to apply mathematical and engineering tools for modeling, simulation, and control design of energy systems and power electronics. The sequence of chapters follows a progressive complexity, but it is possible to change the order or skip material in order to customize a sequence that best fits a combination of the fundamental topics (power electronics, power systems, distributed generation, and renewable energy). Most of chapters are centered on a laboratory project as an example, but some chapters are more discursive with practical explanations of how to model a diversity of electrical engineering systems.

This book follows the approach of problem-based learning strategies, with some project-based learning methodologies. Each chapter has a brief introduction on the theoretical background, a description of the problems to be solved, and objectives to be achieved. Block diagrams, electrical circuits, mathematical analysis, or computer code are also discussed. A solution is presented for the problems or the approach of proposed projects. Each chapter helps the reader to understand the theory, modeling, and computational issues approached in that chapter, with suggestions for further studies, work on possible problems, and even to conduct some experimental work.

1.1 FUNDAMENTALS OF STATE-SPACE-BASED MODELING

Most of the electrical systems consisting of a lumped linear network are causal. They can be written in state-space form such as

$$\dot{x}(t) = Ax(t) + Bu(t) \qquad (1.4)$$

$$y(t) = Cx(t) + Du(t) \qquad (1.5)$$

Such set of first-order differential equations is defined as the state-space equation of the system where $x(t)$ is the state vector, $u(t)$ is the input vector, and $y(t)$ is the output. The second equation is referred to as the output equation. The output is considered a linear combination of states and inputs. The matrix A is called the state matrix, B is the input matrix, C is the matrix of combination of states contributing for the output, and D is the direct transition matrix. One advantage of the state-space formulation is that it is suitable for either analog- or digital-based modeling, control methodology, or mathematical treatment. In addition, the state-space method can be extended to nonlinear systems. State equations can be obtained from a higher-order nth-order differential equation and sometimes from the system model by identifying the appropriate state variables (usually the variables related to energy storage in the system). Suppose an nth-order linear plant model is described by the following differential equation:

$$\frac{d^n y}{dt^n} + a_{n-1} \frac{d^{n-1} y}{dt^{n-1}} + \cdots + a_1 \frac{dy}{dt} + a_0 y = u(t) \tag{1.6}$$

where $y(t)$ is the plant output and $u(t)$ is the plant input. A state model for this system is not unique, and the formulation depends on the choice of a set of state variables. In order to convert this high-order differential equation into state space, we can define the set of state variables, (referred as phase variables) as

$$x_1 = y, \quad x_2 = \dot{y}, \quad x_3 = \ddot{y}, \ldots, \quad x_n = y^{n-1} \tag{1.7}$$

Taking the derivatives, one have
$\dot{x}_1 = x_2, \dot{x}_2 = x_3, \dot{x}_3 = x_4, \ldots$, and after rearranging Equation 1.6, $\dot{x}_n$ is given by

$$\dot{x}_n = -a_0 x_1 - a_1 x_2 - \cdots - a_{n-1} x_n + u(t) \tag{1.8}$$

which can be written in matrix form:

$$\begin{bmatrix} \dot{x}_1 \\ \dot{x}_2 \\ \vdots \\ \dot{x}_{n-1} \\ \dot{x}_n \end{bmatrix} = \begin{bmatrix} 0 & 1 & 0 & \cdots & 0 \\ 0 & 0 & 1 & \cdots & 0 \\ \vdots & \vdots & \vdots & \vdots & \vdots \\ 0 & 0 & 0 & \vdots & 1 \\ -a_0 & -a_1 & -a_2 & \cdots & -a_{n-1} \end{bmatrix} \times \begin{bmatrix} x_1 \\ x_2 \\ \vdots \\ x_{n-1} \\ x_n \end{bmatrix} + \begin{bmatrix} 0 \\ 0 \\ \vdots \\ 0 \\ 1 \end{bmatrix} u(t) \tag{1.9}$$

where the output equation is

$$y = \begin{bmatrix} 1 & 0 & 0 & \cdots & 0 \end{bmatrix} \begin{bmatrix} x_1 \\ x_2 \\ \vdots \\ x_{n-1} \\ x_n \end{bmatrix} \tag{1.10}$$

For example, consider the following differential equation:

$$2\frac{d^3 y}{dt^3} + 4\frac{d^2 y}{dt^2} + 6\frac{dy}{dt} + 8y = 10u(t) \tag{1.11}$$

The coefficient for the highest order derivative must be equal to one, so

$$\frac{d^3 y}{dt^3} + 2\frac{d^2 y}{dt^2} + 3\frac{dy}{dt} + 4y = 5u(t) \tag{1.12}$$

This is a third-order equation, so we can define three state variables as follows:

$$x_1 = y, \quad x_2 = \dot{y}, \quad x_3 = \ddot{y} \tag{1.13}$$

and their derivatives are

$$\dot{x}_1 = x_2, \quad \dot{x}_2 = x_3, \quad \text{and} \quad \dot{x}_3 = -4x_1 - 3x_2 - 2x_3 + 5u(t) \tag{1.14}$$

In matrix form:

$$\begin{bmatrix} \dot{x}_1 \\ \dot{x}_2 \\ \dot{x}_3 \end{bmatrix} = \begin{bmatrix} 0 & 1 & 0 \\ 0 & 0 & 1 \\ -4 & -3 & -2 \end{bmatrix} \begin{bmatrix} x_1 \\ x_2 \\ x_3 \end{bmatrix} + \begin{bmatrix} 0 \\ 0 \\ 5 \end{bmatrix} u(t) \tag{1.15}$$

$$y = \begin{bmatrix} 1 & 0 & 0 \end{bmatrix} \begin{bmatrix} x_1 \\ x_2 \\ x_3 \end{bmatrix} \tag{1.16}$$

A state-space formulation allows mathematical implementation with ODE solvers (which can be computed by MATLAB) and supports the definition of a block diagram for signal modeling simulators (such as Simulink).

1.2 EXAMPLE OF MODELING AN ELECTRICAL NETWORK

State variables are directly related to the energy storage elements of a system, and the ODEs can be derived from nodal or mesh analysis. However, the number of independent state variables depends if there is no loop containing only capacitors and voltage sources and there is no cut-set containing only inductive and current sources. In general, if there are n_C loops of all capacitors and voltage sources, and n_L cut-sets of all inductors and current sources, the number of state variables becomes

$$n = e_L + e_C - n_C - n_L \tag{1.17}$$

where

e_L = number of inductors
e_C = number of capacitors
n_C = number of all capacitive and voltage source loops
n_L = number of all inductive and current source cut-sets

Suppose an electrical circuit, as indicated in Figure 1.1, solve for the equations that define the circuit (using Thévenin) and find the state-space formulation.

Defining the state variables as the current through the inductor and the voltage across the capacitors, we will need eventually first-order equations with $\frac{di_L}{dt}$, $\frac{dv_{C_1}}{dt}$, and $\frac{dv_{C_2}}{dt}$ isolated on the left-hand side of their equations. Initially, write two node equations containing capacitors and a loop equation containing the inductor. The state variables are i_L, v_{c_1}, and v_{c_2}.[1] The KCL equations are

$$0.25\frac{dv_{c_1}}{dt} + i_L + \frac{v_{c_1} - v_i}{4} = 0 \Rightarrow \dot{v}_{c_1} = -v_{c_1} - 4i_L + v_i \tag{1.18}$$

$$0.5\frac{dv_{c_2}}{dt} - i_L + \frac{v_{c_2}}{1} - i_s = 0 \Rightarrow \dot{v}_{c_2} = 2i_L - 2v_{c_2} + 2i_s \tag{1.19}$$

while a KVL loop equation is

$$2\frac{di_L}{dt} + v_{c_2} - v_{c_1} = 0 \Rightarrow \dot{i}_L = 0.5v_{c_1} - 0.5v_{c_2} \tag{1.20}$$

The electrical circuit of Figure 1.1 can be represented by the following state-space equation:

$$\begin{bmatrix} \dot{v}_{c_1} \\ \dot{v}_{c_2} \\ \dot{i}_L \end{bmatrix} = \begin{bmatrix} -1 & 0 & -4 \\ 0 & -2 & 2 \\ 0.5 & -0.5 & 0 \end{bmatrix} \begin{bmatrix} v_{c_1} \\ v_{c_2} \\ i_L \end{bmatrix} + \begin{bmatrix} 1 & 0 \\ 0 & 2 \\ 0 & 0 \end{bmatrix} \begin{bmatrix} v_i \\ i_s \end{bmatrix} \tag{1.21}$$

The state-space formulation aids the composition of a simulation block diagram that can be constructed to model the given differential equations. The basic element of the simulation diagram is the integrator. Assuming a general variable terminology, instead of circuit parameters, say $v_{c_1} = x_1$, $v_{c_2} = x_2$, and $i_L = x_3$, the following matrix equation represents the circuit:

$$\begin{bmatrix} \dot{x}_1 \\ \dot{x}_2 \\ \dot{x}_3 \end{bmatrix} = \begin{bmatrix} -1 & 0 & -4 \\ 0 & 0 & 1 \\ 0.5 & -0.5 & 0 \end{bmatrix} \begin{bmatrix} x_1 \\ x_2 \\ x_3 \end{bmatrix} + \begin{bmatrix} 1 & 0 \\ 0 & 2 \\ 0 & 0 \end{bmatrix} \begin{bmatrix} v_i \\ i_s \end{bmatrix} \tag{1.22}$$

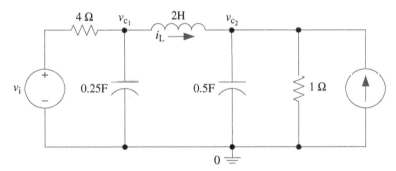

FIGURE 1.1 Electrical circuit with a voltage source and a current source.

[1] The lowercase variables, such as "v_o" and "v_i" are instantaneous values, for example, they change in time, while uppercase variables such as "V_o" and "V_i" are "feature values," such as peak, RMS, average, and they do not change in time.

For each state variable (Figure 1.2), we associate an integrator, $x_1 = \int \dot{x}_1 dt$, such as: It is important to emphasize that the symbol $\dfrac{1}{s}$ is used to represent integration, because the Laplace transform is very often the technique used to arrive to the simulation block diagram. However, the simulation diagram is indeed a time domain representation, and it should be used in circuit simulators where the appropriate numerical solver is selected, with proper selection of time step and tolerance. The number of integrators is equal to the number of state variables. In the state equation of Figure 1.1, there are three integrators with three state variables assigned to the output of each integrator. The last equation is represented via a summing point and connecting the appropriate feedback paths. The output equation should be completed by tapping the right variables with their corresponding multiplication factors and appropriate scaling function. Figure 1.3 shows the simulation block diagram that represents the electrical circuit of Figure 1.1.

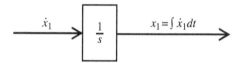

FIGURE 1.2   Each integrator is associated with each state variable.

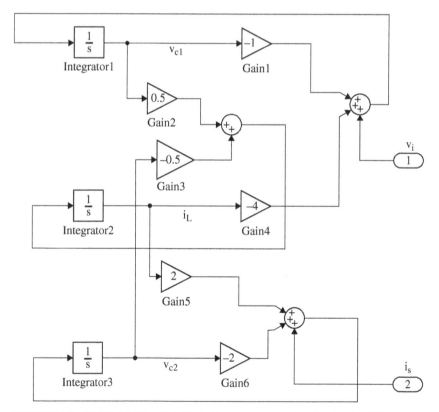

FIGURE 1.3   MATLAB/Simulink block diagram representing the circuit of Figure 1.1.

1.3   TRANSFER FUNCTION

A transfer function of a linear, time-invariant system is defined as the ratio of the Laplace transform of the output (response function), $Y(s) = \mathcal{L}\{y(t)\}$, to the Laplace transform of the input (driving function, or control variable) $X(s) = \mathcal{L}\{x(t)\}$, under the assumption that all initial conditions are zero (Figure 1.4).

It is important in a general system to define which variable is the input, and usually that is associated to a variable that will control the output. A transfer function has the following properties:

1. The transfer function is defined only for linear time-invariant systems.
2. The transfer function between a pair of input and output variables is the ratio of the Laplace transform of the output to the Laplace transform of the input.
3. All initial conditions of the system are set to zero.
4. The transfer function is independent of the input of the system.

A dynamic system can be described by the following time-invariant transfer function related to the input/output system response:

$$G(s) = \frac{Y(s)}{X(s)} \tag{1.23}$$

where the roots of $X(s)$ are called poles of the system, and the roots of $Y(s)$ are called zeros of the system, and the system is considered to have a higher-order polynomial at the denominator, that is, $n > m$:

$$\frac{Y(s)}{X(s)} = \frac{b_0 s^m + b_{m-1} s^{m-1} + \cdots + b_1 s + b_0}{a_0 s^n + a_{n-1} s^{n-1} + \cdots + a_1 s + a_0} \tag{1.24}$$

By setting the denominator function to zero, we obtain what is referred to be the characteristic equation:

$$a_0 s^n + a_{n-1} s^{n-1} + \cdots + a_1 s + a_0 = 0 \tag{1.25}$$

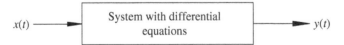

Taking the Laplace transform with zero initial conditions

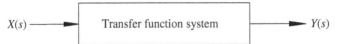

FIGURE 1.4   Laplace transform transforms function derivation for a general system.

The stability of a linear SISO system is completely governed by the roots of their characteristic equation. In order to derive the transfer function of a system, we use the following procedures:

1. Develop the differential equation for a system using the physical laws, for example, Newton's laws, Kirchhoff's laws, power and energy balance, mass flow balance, entropy, and so on. The differential equation should be related of a driving or control variable to the output.
2. Take the Laplace transform of the differential equation under the zero initial conditions.
3. Take the ratio of the output $Y(s)$ to the input $U(s)$. This ratio is the transfer function.

There are experimental procedures that can be used for such transfer function evaluation as well.

1.3.1   State Space to Transfer Function Conversion

Consider the state and output Equations 1.4 and 1.5: $\dot{x}(t) = Ax(t) + Bu(t)$, $y(t) = Cx(t) + Du(t)$
Taking the Laplace transform,

$$sX(s) = AX(s) + BU(s) \Rightarrow [sI - A]X(s) = BU(s) \qquad (1.26)$$

$$Y(s) = CX(s) + DU(s) \qquad (1.27)$$

Substituting for $X(s)$ in the second equation previously, we obtain

$$Y(s) = C[sI - A]^{-1} BU(s) + DU(s) \Rightarrow \frac{Y(s)}{U(s)} = C[sI - A]^{-1} B + D \qquad (1.28)$$

MATLAB has a function $[\text{num}, \text{den}] = \text{ss2tf}(A, B, C, D, i)$ that converts the state equation to a transfer function. The following state-space equations provide an example:

$$\begin{bmatrix} \dot{x}_1 \\ \dot{x}_2 \end{bmatrix} = \begin{bmatrix} 0 & 1 \\ -6 & -5 \end{bmatrix} \begin{bmatrix} x_1 \\ x_2 \end{bmatrix} + \begin{bmatrix} 0 \\ 1 \end{bmatrix} u(t) \qquad (1.29)$$

$$y = \begin{bmatrix} 8 & 1 \end{bmatrix} \begin{bmatrix} x_1 \\ x_2 \end{bmatrix} \qquad (1.30)$$

The system transfer function can be calculated using the matrix definition:

$$[SI - A] = \begin{bmatrix} s & -1 \\ 6 & s+5 \end{bmatrix} \Rightarrow \Phi(s) = [SI - A]^{-1} = \frac{\begin{bmatrix} s+5 & 1 \\ -6 & s \end{bmatrix}}{s^2 + 5s + 6} \qquad (1.31)$$

$$G(s) = C[SI - A]^{-1} B = \begin{bmatrix} 8 & 1 \end{bmatrix} \frac{\begin{bmatrix} s+5 & 1 \\ -6 & s \end{bmatrix}\begin{bmatrix} 0 \\ 1 \end{bmatrix}}{s^2 + 5s + 6} = \frac{\begin{bmatrix} 8 & 1 \end{bmatrix}\begin{bmatrix} 1 \\ s \end{bmatrix}}{s^2 + 5s + 6} = \frac{8+s}{s^2 + 5s + 6} \qquad (1.32)$$

Therefore,

$$G(s) = \frac{s+8}{s^2 + 5s + 6} \qquad (1.33)$$

Another example that will be solved with MATLAB might be described by the following state-space equations:

$$\begin{bmatrix} \dot{x}_1 \\ \dot{x}_2 \\ \dot{x}_3 \end{bmatrix} = \begin{bmatrix} 0 & 1 & 0 \\ 0 & 0 & 1 \\ -1 & -2 & -3 \end{bmatrix} \begin{bmatrix} x_1 \\ x_2 \\ x_3 \end{bmatrix} + \begin{bmatrix} 10 \\ 0 \\ 0 \end{bmatrix} u(t) \qquad (1.34)$$

$$y = \begin{bmatrix} 1 & 0 & 0 \end{bmatrix} \begin{bmatrix} x_1 \\ x_2 \\ x_3 \end{bmatrix} \qquad (1.35)$$

Find the transfer function $G(s) = \dfrac{Y(s)}{U(s)}$.

The following statements:

$$A = \begin{bmatrix} 0 & 1 & 0; & 0 & 0 & 1; & -1 & -2 & -3 \end{bmatrix}; \quad B = \begin{bmatrix} 1 & 0; & 0; & 0 \end{bmatrix};$$

$$C = \begin{bmatrix} 1 & 0 & 0 \end{bmatrix}; \quad D = \begin{bmatrix} 0 \end{bmatrix};$$

$$[num, den] = ss2tf(A, B, C, D, 1)$$

$$G = tf(num, den)$$

results in

num =
   0.0000 10.0000 30.0000 20.0000
den =
   1.0000 3.0000 2.0000 1.0000

and the transfer function computed by the MATLAB function is

$$10s^2 + 30s + 20$$

$$- - - - - - - - - - - -$$

$$s^3 + 3s^2 + 2s + 1$$

Using $[z, p] = ss2tf(A, B, C, D, 1)$, MATLAB will convert the state equation to a transfer function in factored form.

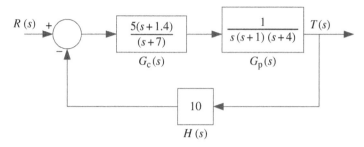

FIGURE 1.5   The feedback function gives the closed-loop transfer function for the closed-loop state-space model.

A closed-loop control is depicted in Figure 1.5. One can use the transfer function to closed-loop function and tf2ss for the closed-loop state-space model of Figure 1.5: The following MATLAB commands:

```
Gc = tf(5*[1 1.4], [1 7])  % transfer function Gc
Gp = tf([1], [1 5 4 0]);   % transfer function Gp
H = 10;
G = series(Gc, Gp)         % connects Gc & Gp in cascade
T = feedback(G, H)         % obtains  the  closed  loop
                             transfer function
[num, den] = tfdata(T, 'v'); % returns num & den as row
                               arrays
[A, B, C, D]=tf2ss(num, den) % returns the A,B, C, D matrices
                             % of the state space model
```

result in the transfer function

$$\frac{5s+7}{s^4+12s^3+39s^2+78s+70}$$

A =
−12	−39	−78	−70
1	0	0	0
0	1	0	0
0	0	1	0

B =
| 1 |
| 0 |
| 0 |
| 0 |

C = [0 0 5 7]

D = 0

1.4   MODELING AND SIMULATION OF ENERGY SYSTEMS AND POWER ELECTRONICS

An analog circuit with linear elements (resistors, inductors, capacitors, voltage and current sources, dependent voltage, and current sources) will mostly be ideal, unless any of the elements may have a nonlinear voltage/current relationship (due to magnetic saturation, temperature variation). However, when a diode or a transistor is used in the circuit, a switching characteristic will introduce challenges in analysis and computer simulation that will make the solution either more difficult or nonlinear, or a numeric solution may need a very robust solver. Switches can be a true open circuit or a short circuit, either depending upon conditions in the circuit (for diodes and thyristors), or they can be commanded by an external controller (transistors, such as IGBTs, Power MOSFETS, and some other ones). Most of time the devices will block a positive voltage when they are off but will not block a negative voltage, for example, a transistor cannot keep withstanding a negative voltage because it may go through breakdown and be damaged. On the other hand, a thyristor (SCRs and related devices) can withstand a certain negative overvoltage without any damage. Therefore, one possible class of designing a power electronic circuit will depend if the switches should be unidirectional voltage blocking (transistors) or bidirectional voltage blocking (SCRs or a more complex connection of transistors and diodes).

In this book, we emphasize the understanding of the structure and circuit topology of the energy conversion and power electronics by assuming that the switches are ideal, that is, there is no any stray capacitance or inductance, no any internal resistances, or any voltage drop when they are on. Also, there is no current leakage when they are off, and the change from on to off and from off to on is instantaneous. Therefore, ideal switches do not show conduction losses (voltage drop multiplied by the switch current) neither switching losses (transition of voltage and current during the switching). If the reader is interested in particularly semiconductor nonideal modeling, he or she should study books and articles in Physics of semiconductors and electronics. The main interest for the power electronics or the power systems designer is to comprehend the functionality of the overall energy conversion, power quality, and control aspects, and the ideal switch is supposed to be the main controllability or lack of controllability that should be accounted.

Because of those ideal switched elements, conditions can occur which are not usually encountered in linear analog circuits without switches. For example, (i) some capacitor voltages or inductor currents might be state variables in one configuration (a certain switch position) and may not be state variables in another one; (ii) the switching may cause capacitor voltages or inductor currents to change instantaneously, producing impulses; this might be an ill-designed circuit, or it may happen in some cases, where the impulsive nature might be a natural condition for a diode to turn on or turn off, when diodes connect branches of circuits with nonzero initial conditions on inductors they are called by freewheeling diodes; (iii) the switching may change the ordinary differential equations from one set to another set of equations, and only techniques of averaging can support linearization of the system.

Consider the simple half-wave rectifier circuit shown in Figure 1.6a, where the configuration for the diode is off (Figure 1.6b) else the diode is conducting, and the equivalent circuit is given by Figure 1.6c. Of course, the diode in this example is ideal, and a nonideal diode model would complicate the understanding of the energy conversion. Nonideal model devices (diodes, transistors, and thyristors) should be employed when the physics of semiconductors and detailed rise and fall voltage and current behavior are important to understand, typically when electronics are designed,

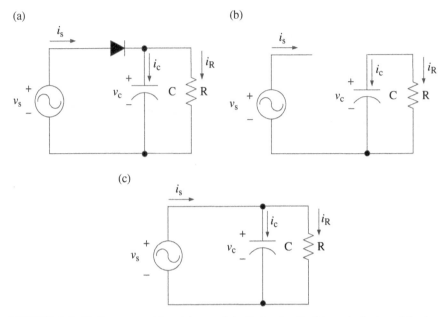

FIGURE 1.6 Half-wave rectifier: (a) complete diode circuit, (b) equivalent model when diode is off, instead of and (c) equivalent model when diode is on.

instead of the power electronics or power systems design. By observing Figure 1.6, we can see that when the diode is open, the circuit is very simple, the input sinusoidal voltage is disconnected, and the capacitor will have an initial voltage and will discharge on the resistor with an exponential response. The equations for this diode off state are $\dfrac{dv_c}{dt} = -\left(\dfrac{1}{RC}\right)v_c$, where the capacitor voltage has initial condition v_c (time when diode just turned off) $= v_c(0_-) = V_{C0-}$, and $i_s(t) = 0$. When the diode is closed, as indicated in Figure 1.6c, the source voltage v_s appears directly across the capacitor, so v_c is no longer a state variable. The equations are now $v_s(t) = v_C(t)$, and the source current is $i_s = \left(\dfrac{1}{R}\right)v_s + C\dfrac{dv_s}{dt}$. In this state, the derivative of the source function needs to be available. In some systems, we may need even higher-order derivatives, and this makes a simple circuit complicated to solve just because a simple diode connects a voltage source with a capacitor in parallel. We can introduce a finite impedance between the source and the capacitor, or make other approximations or more realistic nonideal analysis, but this simple problem demonstrates how the insertion of only one diode can make the system highly nonlinear. This derivative actually shows that any voltage transient at the input will produce impulsive current through the capacitor, which might be harmful in real systems.

A classical problem studied in introduction to electrical circuits in a physics course is depicted in Figure 1.7a, where two capacitors are initially charged with initial voltages, such as V_{C1-} and V_{C2-}. Immediately, after closing, these voltages must be the same, and the new voltage is such that charge is conserved, or

$$V_{C+} = \frac{1}{(C_1 + C_2)} \left[C_1 V_{C1-} + C_2 V_{C2-} \right] \tag{1.36}$$

During switching, there must have an impulsive current through the switch to redistribute the charge. In a real-life situation, the switch could be damaged by a very high current, just limited by the internal resistance of the switch. However, in some cases, it might be necessary to study a system like this, and one technique to handle this situation, that is, capacitors in parallel or inductors in series, is to add to each capacitor a parallel current course and to each inductor a series voltage source. The values of these sources are zero, except during switching when they provide impulses. Just before switching, the charge in all capacitors and the flux in all inductors are calculated, the switches are then set to their next positions, the impulsive sources are applied, and the new charge distribution on the capacitors and flux distribution on the inductors are calculated. For the capacitor circuit considered in Figure 1.7a, two impulse sources are added to apply the charge before switching to an effective capacitor, as indicated in Figure 1.7b.

Figure 1.8a shows a typical boost converter, also called as a flyback circuit. If you are familiar with this circuit, you will know that when S is "on" the inductor has voltage applied, and energy is stored with a linear rise of the inductor current, and by assuming that the output voltage is higher the diode is "off." In this circuit, an impulse

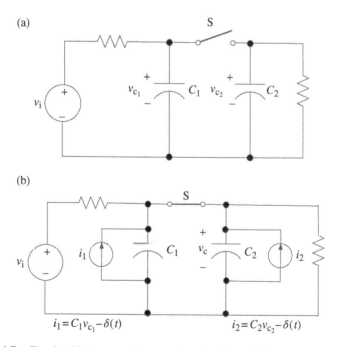

FIGURE 1.7 Circuit with two capacitors connected with a switch: (a) two capacitors with initial voltages before switching and (b) two capacitors with added impulsive sources after switching to make possible total charge balance.

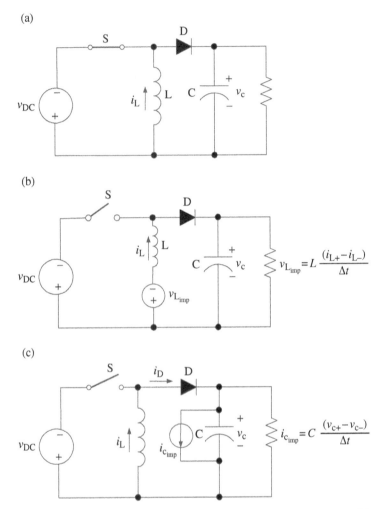

FIGURE 1.8 Impulsive voltage and current behavior: (a) boost converter, (b) voltage impulse produced by inductor, and (c) current impulse produced by capacitor.

produced by the inductor current will impose the diode to change its state from "off" to "on." When the main switch S opens, there is a tendency of the inductor current to go to zero, and a voltage impulse is produced whose magnitude is L times the change in current divided by some small Δ*t* (which can be user specified).

Figure 1.8b shows the voltage impulse produced by the inductor, and Figure 1.8c shows the current impulse produced by the capacitor. The modeling and simulation of power electronic circuits with several switches must consider this kind of impulsive currents and impulsive voltages, balance of charges, and balance of fluxes, and very good and robust mathematical solution of differential equations must be incorporated in the simulators. There are several software packages nowadays for circuit modeling, but this textbook concentrates on PSIM and Power Systems Toolbox. We also recommend PLECS, CASPOC, and SIMPLORER—but we cannot cover all those tools in this book. There are also important object-oriented modeling techniques and languages (such as MODELICA AND DYMOLA). The reader is advised to use this book as a guide, understand our philosophy in teaching, and probably use his or her own software for developing his or her own modeling and simulation for education and instruction in his or her school.

Modeling and simulation of electric circuits started with the ECAP program, which was developed at IBM. ECAP was the first general program for solving time-varying circuit equations. It was useful for different disciplines in electrical engineering requiring different methods for modeling and simulation. Electronic circuit simulations became very popular with the development of Simulation Program with Integrated Circuit Emphasis (SPICE). With computer becoming more powerful for computation and more popular during the 1970s and 1980s, several mathematical equation-oriented solving programs allowed block diagram models or modeling languages, for example, CSMP, TUTSIM, ACSL (or ACSLx), SIMNON, MATRIXx, DYMOLA, MODELICA, VISSIM, and SIMULINK (there are several other ones).

Modeling and simulation methods for power electronics, drive systems, energy conversion, and power electronics developed stronger and with further capabilities in the beginning of the twenty-first century. In addition, there have been mathematical developments of defining ODEs caused by switches in the power electronic conversion circuit (introducing a causal nonlinear relation) and state-space averaging, that is, modeling for the linearization of switched-mode power supplies and inverters for control analysis and design. Simulation is all based on numerically solution of nonlinear state equations, where independent storage elements, such as inductors and capacitors, are described by differential equations.

Because of the differences between the various models for circuits, digital controllers, analog controllers, and components, a multilevel approach can combine the various models such as a circuit model, a block diagram, and even computer program instructions. The combination of these models is a hybrid mathematical model. Numerical solving of this mathematical model reveals the time responses. With numerical methods, time responses can be calculated. If a mathematical model can describe the behavior with linear mathematical relations and constant parameters, the frequency domain can also be used. Using numerical methods, a frequency response can be calculated.

The complexity of the mathematical model is not necessarily related to the complexity of the model of the component. A simple model can contain a (nonlinear) causal mathematical relation, and therefore a DAE has to be used for the mathematical model. On the other side, a more detailed model can be described by ODEs if it does not contain any causal relations. All the nonlinear mathematical relations describing components for energy systems and power electronic systems are functions of time and/or of other variables. These mathematical relations can be formulated as a DAE, making a description by DAEs more general than any other mathematical modeling approach.

The SPICE program is based on the modified nodal analysis (MNA) method, and several other similar circuit-oriented simulators became available in the last few years, such as PSIM, Simplorer, CASPOC, and Power Systems Toolbox. The parameters of the matrix A are dependent on the variables and state variables in the vector $x(t)$. Vector $b(t)$ stores the values of the independent sources. The time derivative of $x(t)$ is replaced by a numerical integration approximation, where the parameter h is the step size of the numerical integration. There are several loops inside the algorithm, one for a recursive Newton–Raphson method. If the convergence of the Newton–Raphson method fails, the step size h is reduced and the ODE is solved again. The zero crossing of currents can be used for evaluation of divergence of the Newton–Raphson method. As a result, the step size is decreased, which can lead to many cycles.

Several different simulators may use different ways to solve a circuit simulation, but in principle, all simulators have a graphical user interface for the definition of the electrical and electronic system, with nodal analysis, constitution of algebraic and differential equations, and use of built-in models for several components, particularly the ones that are more complex with smaller step-size requirements. The electrical circuit will have nonlinear causal mathematical relations with DAE, and then MNA will be used for organizing the system. For a block diagram-oriented simulator, the system is based on nonlinear causal mathematical relations with ODE.

The last few years made possible the total integration of hybrid systems, discrete event, nonlinear, interaction with real-time control, the use of advanced signal processing and artificial intelligence-based techniques. This textbook will explore only the fundamentals of modeling power electronics and interfacing energy conversion systems. The objective in this book is to train undergraduate and graduate students to understand the principles of modeling and to learn the foundation of computer modeling-based design.

1.5 SUGGESTED PROBLEMS

1. A water heater has a resistance to increase the temperature of the water by allowing current to flow through the resistor. A switch (transistor) will turn on and off a DC voltage V_{DC} to modulate the average voltage applied to the resistor (chopper circuit), that is, a PWM circuit with command control $\delta(t)$, where $0 \leq \delta(t) \leq 1$ will impose the turn on and turn off of the switch, for commanding the on time, that is, $\delta = T_{ON} / T$. Suppose there is no inductance on the heating resistance and that the system can be simplified to an equivalent circuit depicted in Figure 1.9.

   For this project you should:

   a) Find all the equations that relate the duty cycle of the PWM to the water temperature increase. Assuming there is no mass change of water during the heating.

   b) Simulate the system using a block diagram and a circuit-oriented approach.

   c) Suppose a proportional control is implemented, that is, a reference set point is compared to the temperature in the water and such difference will be

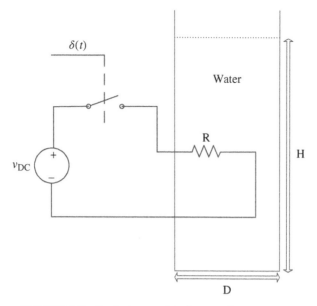

FIGURE 1.9 Equivalent model of a water heater system.

scaled by a gain and applied as the PWM duty cycle command. Simulate this closed-loop system and make sure that is fully operational and stable.

d) Suppose a solar system provides water inlet with a given temperature, and the heater will have to provide only the differential energy to raise the temperature. Calculate the electricity cost for a total electric power system and a hybrid solar/electric water heater.

e) Assume there is mass change during the heating, that is, there is an inlet of water at temperature T_{IN} and a water outlet of water at T_{OUT}. The water height may not change more than 30% from maximum capacity during, that is, the height of water is minimum of 0.7H. Repeat steps (a) to (d) mentioned previously, considering any other constraints or conditions required for your design. Propose an equation-based block diagram simulation for an open loop as well as a closed-loop system.

f) Since the water takes a long time to heat up and the resistor is usually a temperature dependent, nonlinear device, incorporate such real-life behavior in this modeling study.

g) Assume there is mass change of water, i.e. there is an inlet of water and an outlet of water at a given flow and re-do the whole system design for such conditions.

2. This initial chapter introduced several computer-based analysis and simulation techniques based on the principles of understanding the algebraic, differential equations of a system, with possible formulation using numeric solution, block diagram, and circuit-based simulation. This problem introduces two high

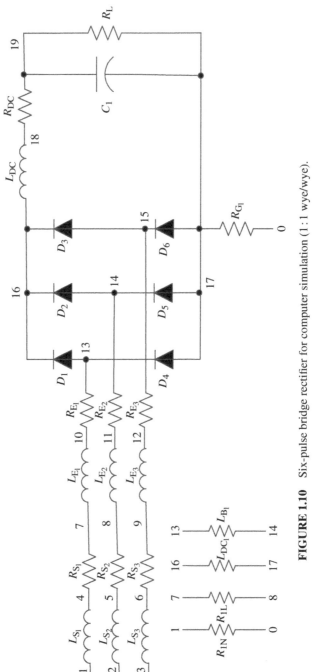

FIGURE 1.10 Six-pulse bridge rectifier for computer simulation (1 : 1 wye/wye).

power rectifier circuits; they are defined as a six-pulse diode bridge converter for three-phase alternating current supply. The one indicated in Figure 1.10 has the typical leakage and stray inductances, but the diodes can be considered ideal. This system is connected to a wye, floating neutral three phase, although resistance R_{G1} might model a current path between the rectifier output to neutral and could also have a wye–wye transformer with some sort of turns ratio. The circuit indicated in Figure 1.11 is also a six-pulse rectifier, but there is a 1:1 delta/wye transformer. There is a reason, related to power quality to be (defined in another chapter), that someone may use either one circuit, or maybe a combination of both, in an industrial installation.

The steps you have to do for this problem are the following:

a) Find the equations and formulation of both systems using all algebraic and differential equations in order to make possible a block diagram system simulation.

b) Simulate both rectifiers using a circuit-oriented simulator (such as PSIM and MATLAB Power Systems Toolbox) and compare with the equation-oriented block diagram approach.

c) Study how a wye/wye and a delta/wye transformer will modify the input utility currents and understand how the current supplied by the utility is different from one case (wye/wye) to another one (delta/wye).

d) Instead of a rectifier with a capacitor and a resistor load, suppose you have a highly inductive load, or maybe a current source at the output of each rectifier. Make adjustments in your simulations to study this case and repeat steps #1, #2, and #3 for highly inductive loads with wye/wye and delta/wye rectifiers

e) Study the Fourier expansion of typical input utility currents for case #4 and understand how 5th and 7th harmonics of one rectifier will counterbalance the other one.

f) Suppose a large power plant will use two rectifiers for two large highly inductive loads, one in wye/wye and another one in delta/wye, and suppose the active power is the same and equally distributed in each rectifier. Simulate a case study in both (i) block diagram and (ii) circuit simulation, and show that your system does not contain 5th and 7th harmonics.

3. Multiple DC output converters are widely used for small power applications, such as in computer power supplies and alarm systems. Figure 1.12 is a simple example of this type of converter, where a PWM control can power several transformer secondary output under different voltage levels, as the one represented in this figure. Explore this circuit to design a $+5\,V_{DC}$ and a $\pm12\,V_{DC}$ voltage sources.

For this project, you will have to:

a) Find all the equations that relate the duty cycle of the PWM to each of three interdependent output voltages.

b) Simulate the system using a block diagram and a circuit-oriented approach.

c) Assuming a pondered reference voltage for all sources, establish the proper duty cycle to keep them as close to this reference under distinct load currents

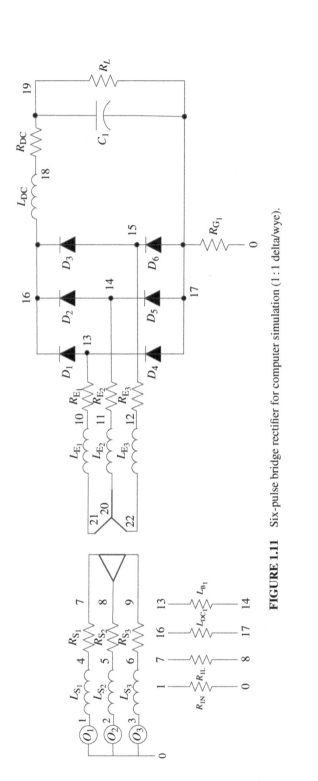

FIGURE 1.11 Six-pulse bridge rectifier for computer simulation (1 : 1 delta/wye).

for each sources. Use a PI control for such difference scaled by a gain and applied the PWM duty cycle command. Simulate this closed-loop system and make sure that is fully operational and stable.

d) Assume the worst case of current unbalance and establish the technical specifications for such multiple voltage source. Repeat steps (a) to (c) mentioned previously, considering any other constraints or conditions required for your design, for example, one case of voltage sensitive load.

e) Propose an equation-based block diagram simulation for an open loop as well as a closed-loop system.

4. Virtual short circuit is a common problem in electrical engineering, for example, zero ground voltage V_g and current I_g in well-balanced three-phase systems and in the differential input of operational amplifiers (Figure 1.13). This is not exactly true when we have generation of harmonics circulating through the lines from nonlinear loads.

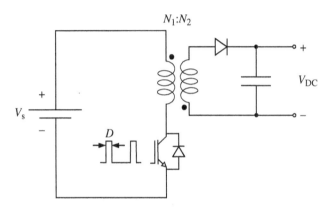

FIGURE 1.12   Basic flyback converter.

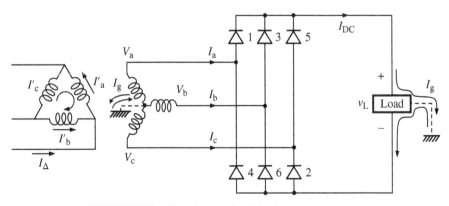

FIGURE 1.13   Virtual short circuit in three-phase systems.

For this project, you will have to:

a) Find all the equations that relate the virtual ground short-circuit currents to the well-balanced three-phase power source.
b) Simulate the system using a block diagram and a circuit-oriented approach.
c) Plot and calculate the RMS currents through the ground connection during the virtual short circuit.
d) Produce a 10% current unbalance in phase "a" and observe the ground current.
e) Repeat steps (a) to (d) mentioned previously, considering the effects of an impedance placed in the ground wire.
f) Consider other constraints or conditions required for your design, for example, overcurrent protection level and minimum-phase voltage level.

5. Figure 1.14 shows a circuit working with current interruption. Make analysis to obtain all the currents just soon after closing the switch at $t=0$, representing the active load modeled by a voltage source active after the switch is ON.
For this project, you will have to:

a) Find all the equations relating active and passive components.
b) Simulate the system using a block diagram and a circuit-oriented approach.
c) Suppose the current source has a current of 0.5 A step at $T = 0.01$s. How would this circuit behave under the point of view of current and voltage surges?
d) Repeat step (c) mentioned previously, considering a voltage step of 10 V step also at $T = 0.01$s.
e) Repeat step (c) mentioned previously, considering any other constraints or conditions required for your design.
f) Propose an equation-based block diagram simulation for an AC current source and an AC voltage source.

Parameter	Value
I_{DC}	2.5 A
V_{DC}	12 V
R_1	10 Ω
R_2	20 Ω
R_3	30 Ω
R_4	40 Ω
C_1	1 μF
C_2	2 μF

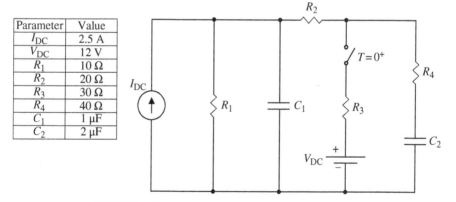

FIGURE 1.14   Switched circuit based on RC responses.

FURTHER READING

BALBANIA, N. and BICKART, T.A., Electrical Network Theory, John Wiley & Sons, New York, 1969.

BANKS, J., CARSON, J. and NELSON, B., Discrete Event System Simulation, Prentice Hall, Upper Saddle River, 1996.

BAUER, P., DUIJSEN, P. and VAN, J., "Large signal and small signal modeling techniques for AC-AC power converters", Proceedings of PCC-Yokohama, IEEE, pp. 520–525, 1993.

HO, C.-W., RUEHLI, A.E. and BRENNAN, P.A., "The modified nodal approach to network analysis", IEEE Transactions on Circuits and Systems, vol. CAS-22, no. 6, pp. 504–509, 1975.

CARNAHAN, B., LUTHER, H.A. and WILKES, J.O., Applied Numerical Methods, John Wiley & Sons, Inc., New York, 1969.

CUK, S. and MIDDLEBROOK, R.D., "A general unified approach to modeling switching dc-dc converters in discontinuous conduction mode", Proceedings PESC 1977, pp. 36–57, 1977.

DIRKMAN, R.J., "The simulation of general circuits containing ideal switches", IEEE 0275-9306/87/0000-0185, 1987 IEEE, pp. 185–194, 1987.

FISHWICK, P., Simulation Model Design and Execution: Building Digital Worlds, Prentice Hall, Englewood Cliffs, 1995.

FUJIMOTO, R.M., Parallel and Distributed Simulation Systems, John Wiley & Sons, New York, 2000.

GHOSH, A. and LEDWICH, G., Power Quality Enhancement Using Custom Power Devices, Kluwer's Power Electronics and Power Systems Series, Kluwer Academic Publishers, Boston, 2002.

JAMES, M.L., SMITH, G.M. and WOLFORD, J.C., Applied Numerical Methods for Digital Computation, University of Nebraska, IEP-A Dun-Donnelley Publisher, New York, 1977.

KELTON, W.D and AVERILL, M., Simulation Modeling and Analysis, McGraw Hill, Boston, 2000.

MIDDLEBROOK, R.D. and ĆUK, S., "A general approach to modelling switching-converter power stages", Proceedings PESC 1976, pp. 18–31, 1976.

MOHAN, N., UNDELAND, T.M. and ROBBINS, W.P., Power Electronics: Converters, Applications and Design, John Wiley & Sons, Inc., New York, 1989.

MORRIS, N.M., Advanced Industrial Electronics, McGraw-Hill Book Co. Ltd., London, ISBN-10: 0070846944, ISBN-13: 978-0070846944, 1984.

NAKAHARA, M. and NINOMIYA, T., "A general computer algorithm for analysis of switching converters", IEEE PESC 1992, pp. 1181–1188, 1992.

NELMS, R.M. and GRIGSBY, L.L., "Simulation of power electronic circuits containing nonlinear inductances using a sampled-data model", APEC-IEEE, CH2853-0/90/0000-0746, pp. 746–749, 1990.

O'MALLEY, J., Theory and Problems of Basic Circuit Analysis, Schaum's Outline Series, McGraw-Hill, Inc., New York, 1982.

RAGAZZINI, J.R., RANDALL, R.H. and RUSSELL, F.A., "Analysis of problems in dynamics by electronic circuits", Proceedings of the IRE, vol. 35, no. 5, pp. 444, 452, 10.1109/JRPROC.1947.232616, May 1947.

2

ANALYSIS OF ELECTRICAL CIRCUITS WITH MESH AND NODAL ANALYSIS

2.1 INTRODUCTION

Network Topology is a branch of electrical circuit theory concerned with equations and theorems required to completely describe an electrical system. In this book, we are mostly concerned with the following two theorems, applicable to a regular circuit (even for very big circuits), which can be represented by a planar graph:

1. Let N = number of nodes in a circuit; then $N-1$ independent nodal equations are required to completely describe the circuit. These equations are obtained by setting the algebraic sum of currents, leaving each of the $N-1$ nodes equal to zero; this is called Kirchhoff's current law, abbreviated as KCL or nodal analysis.
2. Let $L = M$ = number of loops or meshes, B = number of branches, N = number of nodes in a circuit; then $L = M = B - N + 1$ independent loop or mesh equations are required to completely describe the circuit. These equations are obtained by setting the algebraic sum of the voltage drops around each of the $L = M = B - N + 1$ loops or meshes equal to zero; this is called Kirchhoff's voltage law, abbreviated as KVL or mesh analysis.

When a circuit is described by their mesh or nodal analysis, a linear system equation will have to be solved either by hand or by numerical analysis, such as $Ax = y$, where y is a vector, A is a square matrix, and x is a vector.

Modeling Power Electronics and Interfacing Energy Conversion Systems, First Edition.
M. Godoy Simões and Felix A. Farret.
© 2017 John Wiley & Sons, Inc. Published 2017 by John Wiley & Sons, Inc.

Depending on the analysis, being performed to be either mesh or nodal, y is a vector of either node voltages or mesh currents, and A is a matrix defined by either impedances or admittances. For physical well-defined systems, this linear system $Ax = y$ has a matrix A where $m = n$, that is, it is a square system with a unique solution. Therefore, the issue is to invert such matrix, either by hand or by computational techniques. The matrix A can be only made of real numbers, for resistive circuits; may have complex numbers, for phasor analysis (sinusoidal steady state); or may have Laplace operators, for transient analysis. For such transient analysis, the matrix A must be solved by symbolic software manipulation or by changing the Laplace-based matrix into state space and solving with ordinary differential equation numerical computation (Chapter 3).

2.2   SOLUTION OF MATRIX EQUATIONS

In order to solve matrix equations, we need to specify whether a pre- or postmultiplication has to be done, because matrices do not obey the commutative law ($AB \neq BA$). Given that A, B, and C are matrices and $AB = C$, where A and B are nonsingular, that is, square and invertible, then it is possible to premultiply both sides of the equation by $(A-1)$ and $(A-1)AB = (A-1)C$. Since A is nonsingular, we have that $(A-1)A = I$, and the matrix B can be isolated as $B = (A-1)C$. This can be applied to the solution of the linear system as $x = (A-1)y$, that is, any algorithm that inverts A will solve this problem. The project in this chapter deals with such matrix computations for electrical circuits.

There is another interesting perspective. Suppose we want to isolate A, so given that B is invertible, then we can postmultiply by $(B-1)$ and $AB(B-1) = C(B-1)$ if B is nonsingular, and we can make sure that $(B-1)B = I$, the matrix A can be isolated to $A = C(B-1)$ if A is an incidence matrix representing a graph or an input/output mapping. That means the observation of input and output matrices (here considered to be C and B) will make possible the computation of the mapping matrix A. Of course, this methodology cannot be used in an electrical circuit where $Ax = y$ and both x and y are vectors, but such computation can be easily performed for other signal-processing problems or in probabilistic approaches (where this matrix would represent a Bayesian probabilistic density function relationship between two sets). A neural network or a fuzzy logic-based system could also approximate A using artificial intelligence-based techniques, but all these algebraic challenges are out of scope of this book [1].

Why is it important to specify whether we are pre- or postmultiplying matrix equations? Either a pre- or a postmultiplication may have a term such as $(B-1)AB$ impossible to simplify. A matrix calculation is not swappable as it is normally done with real or complex numbers. This kind of matrix computation given by $(B-1)AB$ is a very important linear algebra consideration used in the study of graph-based problems, control analysis, and signal processing. The studies related to such matrix equation can be conducted in advanced control and optimization topics [2, 3].

2.3 LABORATORY PROJECT: MESH AND NODAL ANALYSIS
OF ELECTRICAL CIRCUITS WITH SUPERPOSITION THEOREM

In this laboratory, you will enhance your understanding of circuit analysis using MATLAB for solving numerically an electrical circuit and also PSIM to find their voltages and currents using a circuit simulation. You will have to calculate the power factor for each power source in the circuit, both computed numerically and also comparing with the circuit simulation. As an example, the following circuit (Figure 2.1) has the parameters:

$$v_o(t) = 169.7\cos(2\pi60t + 0.25)(V)$$

$$i_x(t) = 1.5\sin(2\pi60t - 0.15)(A)$$

$R_1 = 10\Omega,\ R_2 = 1.5\Omega,\ R_3 = 5\Omega,\ C_1 = 200\mu F,\ C_2 = 400\mu F,$ and $L_1 = 750\,mh$

1. Using nodal and mesh analysis and phasor analysis, find the current and voltage at each element in their phasor form plus their correspondent time-domain response.

2. What is the phasor form and correspondent time response for current flowing out of the voltage source?

3. Plot input current and output voltage using a theoretical analysis.

4. What is the power factor seen by each source?

5. Do the same using MATLAB.

6. Prepare a very good documentation of your MATLAB script (.M), showing how the nodal analysis and the mesh analysis where implemented in MATLAB. Compare all the theoretical versus the computational results.

7. Assume that the current source is $i_x(t) = -1.5\sin(2\pi60t - 0.15)$ (A). Compare the instantaneous power, active power, reactive power, and power factor for the previous case (positive current source) versus such negative current source. Explain what you observe.

8. Suppose you want to calculate the power across the resistance in series with the inductance (P_L). Give a theoretical framework and a possible simulation implementation.

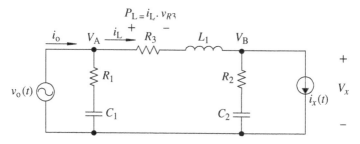

FIGURE 2.1 Electrical circuit for mesh and nodal analysis.

A general procedure for using a nodal analysis method to solve electric circuit problems supports a matrix system by applying KCL at the major nodes in an electric circuit. If the system size is small (two or three nodes), algebra and Cramer's rule can be used to solve the unknown major node voltages. Large systems can be solved using mathematical software, such as MATLAB or many others. After the node voltages are solved, regular circuit analysis methods (Ohm's law, voltage and current divider principles, and so on) can then be used to find any circuit variable or calculation entity is required. The nodal analysis algorithm can be summarized in the following steps:

1. Choose a reference node. (rule of thumb: take the node with most branches connecting to it), that is, usually called "ground," but it is not really necessary to be the physical ground of the circuit—it can be any convenient node; floating circuits may require a good choice for the reference node.
2. Identify and number major nodes and then label them.
3. Apply KCL to identified major nodes and formulate circuit equations.
4. If you have voltage sources at the nodes, they are already the nodal voltages; if you have dependent sources, you have to write extra equations for those dependent sources.
5. Create a matrix system from the KCL equations obtained.
6. Solve the matrix for unknown node voltages by using Cramer's Rule or the Gaussian method, or solve it numerically.
7. Use the solution, that is, node voltages, to solve for the desired circuit variables.

For the circuit depicted in Figure 2.1, the impressed voltage across node A and ground is $v_0(t) = 169.7\cos(2\pi 60t + 0.25)$, so the angle in degree is $0.25\frac{180}{\pi} = 14.324°$. Therefore, the phasor notation using the phasor magnitude in RMS is $\overline{V_0} = 120\angle 14.324°$. In order to convert instantaneous sine and cosine forms to one another, it is important to remind these simple trigonometric identities: $\sin(\omega t + \alpha) = \cos(\omega t + \alpha - 90°)$ and $\cos(\omega t + \alpha) = -\sin(\omega t + \alpha - 90°)$.

The current source in the circuit is defined as $i_x(t) = 1.5\sin(2\pi 60t - 0.15)$. A phasor is assumed to have phase $0°$ for an instantaneous variable such as $\cos(\omega t \pm 0°)$. So, for the current i_x notated with $\sin(\ )$, it is required to subtract $90°$ from its phase shift, that is, $\phi_i = -\left(0.15\frac{180}{\pi}\right) - 90° = -98.6°$. Therefore, the phasor notation for the current source $i_x(t)$ is $\overline{I_x} = 1.06\angle -98.6°$. The nodal analysis, assuming node voltages to be $\overline{V_A}$ and $\overline{V_B}$ gives for two nodes, two equations I and II:

- Node #1:

$$\overline{V_A} = \overline{V_0} = 120\angle 14.324°$$

- Node #2:

$$\frac{\overline{V_B}}{1.5-j6.63}+\frac{\overline{V_B}-\overline{V_A}}{5+j282.743}+\overline{I}_x=0$$

In matrix notation, the equation becomes a linear equation of the form $Ax = y$:

$$\begin{bmatrix} 1 & 0 \\ \left(\dfrac{1}{5+j282.743}\right) & \left(\dfrac{-1}{1.5-j6.63}+\dfrac{-1}{5+j282.743}\right) \end{bmatrix}\begin{bmatrix} \overline{V_A} \\ \overline{V_B} \end{bmatrix}=\begin{bmatrix} 120\angle14.324° \\ 1.06\angle-98.6° \end{bmatrix}$$

The matrix A has admittance components. For such simple 2×2 matrix-based system, it is possible to solve by hand or numerically. With some algebraic work, one can find that $\overline{V_B}=4.83\angle-9.878°$. The current flowing out of the voltage source $(\overline{I}_0)$, which can be calculated by adding the current flowing from the node A to ground $\dfrac{\overline{V_A}}{R_1+\dfrac{1}{j\omega C_1}}$ plus the current flowing from node A to node B, which is $\dfrac{\overline{V_A}-\overline{V_B}}{R_3+j\omega L_1}$, resulting in $\overline{I}_0=6.91\angle65.18°$.

The voltage and current of each element of this circuit can be calculated using a MATLAB script with the complete computation using the algebraic solution instead of a matrix-based inversion.

```
%%Nodal analysis for circuit of figure 2.1%%
%parameters definition
clear all
R1=10;
R2=1.5;
R3=5;
C1=200E-6;
C2=400E-6;
L1=750E-3;
A=169.7;
theta=0.25;
V0 = 169.7*exp(j*0.25);
Ix=1.5*exp(j*1.72089)
w=2*pi*60;
VB=(1/(1/(R2+1/(1i*w*C2))+1/(R3+1i*w*L1))*(1/
  (R3+1i*w*L1)*V0-Ix));
I2=(1/(R1+1/(1i*w*C1))+1/(R3+1i*w*L1))*V0-1/
  (R3+1i*w*L1)*VB;

B = abs(VB);
phi = angle(VB);
```

```
T = 2*pi/w;
tf = 2*T; N = 100; dt = tf/N;
t = 0 : dt : tf;
%------------------------------------------------------
%            Plot V0 and VB.
%------------------------------------------------------
for k = 1 : 101
    V0(k) = A * cos(w * t(k) + theta);
    VB(k) = B * cos(w * t(k) + phi);
end
figure;plot (t, V0, t, VB)
%%end of Matlab script ----------------------------------
```

Another possible way to perform analysis of electrical circuits is using a generalized mesh analysis. The aim of such algorithm was to develop a matrix system, from equations applying KVL around loops or meshes, in an electric circuit. Such methodology is summarized as:

1. Use the conventional current flow (passive notation) for defining positive voltage drop.
2. Identify and number loops or meshes (usually a few ones can be done by hand); each mesh current is assumed to be clockwise.
3. Apply KVL to meshes/loops and formulate circuit equations.
4. If you have current sources in the loops, they are already the mesh currents; if you have dependent sources, you have to write extra equations for those dependent sources.
5. Create a matrix system from KVL equations obtained; solve the matrix either algebraically by hand (with Cramer's rule or the Gaussian method) or by computer-aided mathematical software (MATLAB); use the solution, that is, mesh currents, to solve for the desired circuit variables.

The circuit depicted in Figure 2.1 has a current source; in order to do a mesh analysis, it is possible in two paths; one is to convert the current source with a parallel impedance into a voltage source with series impedance, such as indicated in Figure 2.2, and then we have two meshes, with two mesh currents to solve (2×2 system).

Another path is to work with the full original circuit (3×3 system), imposing one mesh current to be $\overline{I}_x$ (solving for steady-state phasor model). Next, this general procedure is described, where the KVL applied for each mesh is as follows:

- **Mesh #1**

$$-\overline{V}_0 + R_1\left(\overline{I}_1 - \overline{I}_2\right) + \frac{1}{j\omega C_1}\left(\overline{I}_1 - \overline{I}_2\right) = 0$$

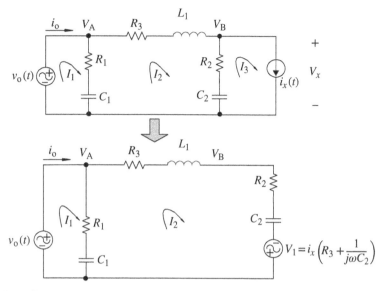

FIGURE 2.2   The current source with a parallel impedance can be converted into a voltage source with series impedance.

- **Mesh #2**

$$-\frac{1}{j\omega C_1}\left(\overline{I}_1-\overline{I}_2\right)-R_1\left(\overline{I}_1-\overline{I}_2\right)+R_3\left(\overline{I}_2\right)+j\omega L_1\left(\overline{I}_2\right)+R_2\left(\overline{I}_2-\overline{I}_3\right)+\frac{1}{j\omega C_2}\left(\overline{I}_2-\overline{I}_3\right)=0,$$

Collecting terms:

$$-\left(\frac{1}{j\omega C_1}+R_1\right)\left(\overline{I}_1\right)+\left(\frac{1}{j\omega C_1}+R_1\right)\left(\overline{I}_2\right)+R_3\left(\overline{I}_2\right)+j\omega L_1\left(\overline{I}_2\right)$$
$$+\left(R_2+\frac{1}{j\omega C_2}\right)\left(\overline{I}_2\right)-\left(R_2+\frac{1}{j\omega C_2}\right)\left(\overline{I}_3\right)=0$$

- **Mesh #3**

$$\overline{I}_3=\overline{I}_x$$

Such linear system has a matrix formulation as $ZI = V$, that is,

$$\begin{bmatrix} \left(R_1+\dfrac{1}{j\omega C_1}\right) & -\left(R_1+\dfrac{1}{j\omega C_1}\right) & 0 \\[2ex] -\left(R_1+\dfrac{1}{j\omega C_1}\right) & \left(R_1+R_2+R_3+\dfrac{1}{j\omega C_1}+\dfrac{1}{j\omega C_2}+j\omega L_1\right) & -\left(R_2+\dfrac{1}{j\omega C_2}\right) \\[2ex] 0 & 0 & 1 \end{bmatrix}\begin{bmatrix} \overline{I}_1 \\[1ex] \overline{I}_2 \\[1ex] \overline{I}_3 \end{bmatrix}=\begin{bmatrix} \overline{V}_0 \\[1ex] 0 \\[1ex] \overline{I}_x \end{bmatrix}$$

where

$$\overline{V}_0 = 120\angle14.324°$$

$$\overline{I}_x = 1.06\angle98.6°$$

```
%% Mesh analysis for circuit of figure 2.1%%
 vo = 169.7*exp(j*0.25)
 ix = 1.5*exp(j*(-0.15-pi/2));
R1 = 10; R2 = 1.5; R3 = 5;
C1 = 200e-6; C2 = 400e-6; L1 = 750e-3;
w=120*pi;
Xc1 = (w*C1); Xc2 = (w*C2); XL= w*L1;
z11 = R1 + 1/(i*Xc1);
z12 = -(R1 + 1/(i*Xc1));
z13 = 0;
z21 = -(R1 + 1/(i*Xc1));
z22 = (R1+R2+R3+1/(i*Xc1)+1/(i*Xc2)+i*XL);
z23 = -(R2 + 1/(i*Xc2));
z31 = 0;
z32 = 0;
z33 = 1;
Z = [z11, z12, z13; z21, z22, z23; z31, z32, z33];
V = [vo;0;ix];
mesh_currents =inv(Z)*V;
I1 = mesh_currents(1,1);
I2 = mesh_currents(2,1);
I3 = mesh_currents(3,1);
vx = (R2 + 1/(i*Xc2))*(I2-I3);
B = abs(I1);
phi = angle(I1);
A=abs(vo);
theta=angle(vo);
Voc=vo-z23*ix

C = abs(I3);
phi1 = angle(I3);
D = abs(vx);
phi2 = angle(vx);
T = 2*pi/w;
tf = 4*T; N = 100; dt = tf/N;
t = 0 : dt : tf;
%--------------------------------------------------
%    Plot Vo and io.
%--------------------------------------------------
```

```
for k = 1 : 101
    V0(k) = A * cos(w * t(k) + theta);
    I0(k) = B * cos(w * t(k) + phi);
end
plot (t, V0,'r',t, I0,'--')

for k = 1 : 101
    IX(k) = C * cos(w * t(k) + phi1);
    Vx(k) = D * cos(w * t(k) + phi2);
end
plot (t, Vx,'r',t, IX,'--')

figure
plot (t, V0,'r',t, I0,'--')
%%end of Matlab script --------------------------------
```

It is possible to improve the previous MATLAB script in order to calculate any circuit voltage, current, or instantaneous power, using basic Ohm's law, as this approach is based on phasor-based steady-state evaluation.

Power factor is the ratio of active power to the apparent power, defined across two terminals, that is, the ratio of average power divided by a denominator composed by the multiplication of the true RMS of voltage and the true RMS of the current:

$$PF = \frac{P}{S} = \frac{\frac{1}{T}\int_t^{t+T} v(t)\cdot i(t)\,dt}{\sqrt{\frac{1}{T}\int_t^{t+T} v^2(t)\,dt}\sqrt{\frac{1}{T}\int_t^{t+T} i^2(t)\,dt}}$$

When the involved voltages and currents are steady-state sinusoids, for a single-phase system, the aforementioned ratio is simplified to PF = $\cos(\varnothing_{vi})$ where $\varnothing_{vi}$ is the phase shift of current in respect to voltage. If current is lagging, that is a lagging power factor situation, whereas if current is leading in respect to the voltage, that is a leading power factor condition.

In the current problem, you can use the MATLAB script to find out the current at the voltage source. The solution is $i_o = 6.91\angle 65.18°$, and the instantaneous waveform can be written as $i_o = 6.91\sqrt{2}\cos(\omega t + 65.18°) = 9.77\cos(\omega t + 65.18°)$. The voltage source was defined initially as $v_o = 169.7\cos(\omega t + 14.324°)$. Therefore, the phase shift is $\varnothing_{vi} = \varnothing_v - \varnothing_i = 14.32° - 65.18°$. The power factor is PF = $\cos(-50.86°) = 0.63$ lagging.

For the current source, it has been originally defined that $i_x = 1.5\sin\left(\omega t - 0.15\frac{180}{\pi}\right)$ $= 1.5\cos(\omega t - 98.6°)$. The voltage across the current source can be found using either the nodal or mesh analysis, that is, $v_x = 4.83\sqrt{2}\cos(\omega t - 9.88°)$. The phase shift is $\varnothing_{vi} = \varnothing_v - \varnothing_i = -9.88° - (-98.6°)$, and the power factor across the current source is PF = $\cos(88.72°) = 0.0223$ leading. Figures 2.3 and 2.4 both show the voltage and current plots for those calculations. Note that when i_x is reversed, more active power is injected into the network.

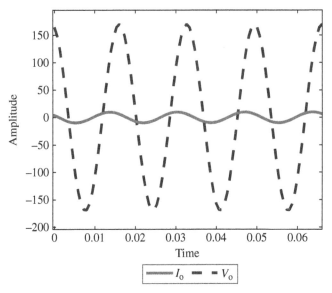

FIGURE 2.3   Voltage source and its current through.

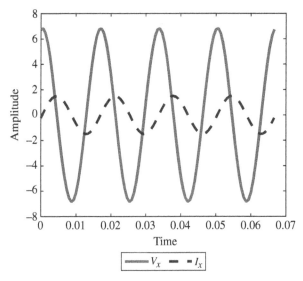

FIGURE 2.4   Current source and its voltage across.

	Normal Current Direction	**Reversed Current Direction**
Power Factor	Voltage source: 0.63 (lagging) Current source: 0.0223 (leading)	Voltage source: 0.6343 (lagging) Current source: 0.34 (leading)
Active Power	Voltage source: 523.56 W Current source: −0.1149 W	Voltage source: 522.6 W Current source: 3.66 W
Reactive Power	Voltage source: −643 VAR Current source: −5.12 VAR	Voltage source: −636.962 VAR Current source: −10.12 VAR

The last item in this project requests to calculate the power dissipated in the resistance in series with the inductance (PL). Analytically it is possible to solve by hand and find the voltage and current across the resistance; it is also possible to write in the MATLAB script that the equations to find the voltage drop across and current through. However, it is more interesting to define the equivalent Thévenin circuit seen by that resistance. The equivalent Thévenin can be calculated algebraically as usually taught in electrical circuits courses. Or it can be resolved through computation by removing the element (the resistance) and measuring the open-circuit voltage across the terminals of the removed element, that is the equivalent Thévenin voltage, V_{th}. By implementing a short circuit and measuring the short-circuit current I_{sc}, the equivalent Thévenin impedance is found by computing $Z_{th} = \dfrac{V_{th}}{I_{sc}}$. We recommend the reader to implement a MATLAB script where you will perform such Thévenin calculations in order to define the equivalent Thévenin model, that is, phasor V_{th} voltage and impedance Z_{th}. The same procedure can conducted using a Simulink block diagram and a PSIM circuit-based simulation.

2.4 SUGGESTED PROBLEMS

1. For the laboratory project discussed in the aforementioned chapter, make the assumption of an applied distorted input voltage (e.g., a square wave, but that could be any other periodic waveform). Study and implement a Fourier expansion to obtain several frequency-tuned equivalent models. Those can be studied using the superposition theorem allowing the evaluation of effects of each harmonic response in the circuit. Implement a MATLAB-based solution for this circuit with distorted voltage input.

2. Model the circuit in the following, which represents an analog FET transistor open-loop circuit (Figure 2.5). Find parameters in a textbook related to electronics. Find the voltage gain and then the power gain (the active power at R_L divided by the input power). Study the system frequency response using a mathematical formulation and comparing it with a MATLAB numerical solution.

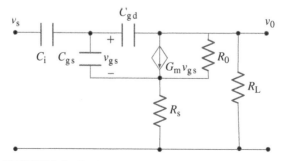

FIGURE 2.5 Equivalent circuit of a field-effect transistor.

3. Model the circuit in the following, which represents the equivalent model for an inverting operational amplifier (Figure 2.6). Find parameters in a textbook related to electronics. Find the voltage gain, $\dfrac{|v_o|}{|v_s|}$, and then study the system frequency response using a mathematical formulation and comparing it with a MATLAB numerical solution.

4. Model the circuit in the following, which represents the per-phase equivalent model for an induction machine (Figure 2.7). Find how the mechanical power can be calculated and plot, using MATLAB, a family of curves of output power related to electrical angular frequency response, slip, and input voltage.

5. Model the photovoltaic cell equivalent model, where I_L represents the current impressed by the solar irradiation, and the diode must be replaced by an appropriate equivalent model (Figure 2.8).

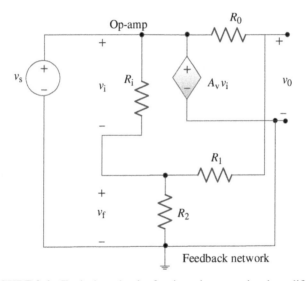

FIGURE 2.6   Equivalent circuit of an inverting operational amplifier.

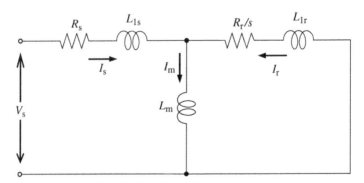

FIGURE 2.7   Induction machine per-phase equivalent model.

6. Model the equivalent circuit in the following to represent a power FET transistor using basic circuit components from MATLAB (Figure 2.9). From literature, select one power MOSFET to attend a load of $I_{C_{max}}$ = 50 A A at 800 V and maximum reverse voltage = 10 V. The other parameters such as the nonlinear resistance limits of the carrier injection modulation R_{mod}, collector resistance R_C, and g_m of the integrated FET and PNP/NPN transistors parameters can be found in commercial data sheets. Then plot a family of output power characteristics for this model relating them to the gate voltage. What would be your critics about this model?

7. A thyristor has a linearized characteristic in the normal operating range as represented in Figure 2.10. Write a MATLAB script to simulate the average power loss characteristics for (a) constant current = 23 A, (b) half-sine wave current with an average value = 18 A, (c) a current level of 39.6 A during a half cycle, and (d) an average current level of 48.5 A during 1/3 cycle. Note: The RMS value of the first current cycle can be used as the nominal value because the direct current is rather low for such case.

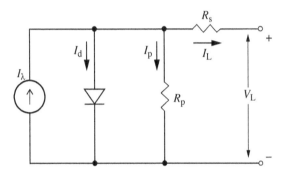

FIGURE 2.8   Photovoltaic cell equivalent model.

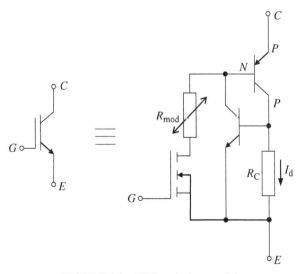

FIGURE 2.9   FET equivalent model.

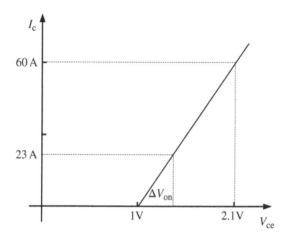

FIGURE 2.10   Thyristor cut-in voltage.

REFERENCES

[1] TSOUKALAS, L.H., UHRIG, R.E. and ZADEH, L.A., Fuzzy and Neural Approaches in Engineering, John Wiley & Sons, New York, ISBN-13: 978-0471160038, ISBN-10: 0471160032, 1996.

[2] WON YOUNG, W., CHUNG, Y., CAO, W., CHUNG, T.S. and MORRIS, J., Applied Numerical Methods Using Matlab, John Wiley & Sons, Hoboken, ISBN: 0-471-69833-4 (cloth), 2005.

[3] CHAPRA, S., Applied Numerical Methods with MATLAB: For Engineers & Scientists, McGraw-Hill Science, New York, ISBN-10: 0073401102, ISBN-13: 978-0073401102, 2011.

FURTHER READING

AYRES, JR, F., Schaum's Theory & Problems of Matrices, Schaum's Publishing Co, New York, 1962.

BRONSON, R. and COSTA, G.B., Differential Equations, 4th edition, Schaum's Outline Series, McGraw-Hill, 2014.

CARNAHAN, B., LUTHER, H.A. and WILKES, J., Applied Numerical Methods. John Wiley & Sons, New York, ISBN-10: 0471135070, ISBN-13: 978-0471135074, 1969.

DESOER, C.A. and KUH, E.S., Basic Circuit Theory, Kōgakusha'-McGraw-Hill, Inc., New York, 1969.

DIXON, C., Linear Algebra, The University Series in Undergraduate Mathematics, Van Nostrand Reinhold Company, New York, ISBN-10: 0442221576, ISBN-13: 978-0442221577, 1971.

GRANVILLE, W.A. and SMITH, P.F., Elements of the Differential and Integral Calculus, revised edition, The Athenæum Press, Boston, 1911.

KREIDER, D.L, KULLER, R.G., OSTBERG, D.R. and PERKINS, F.W., An Introduction to Linear Analysis, Addison-Wesley Publishing Company Inc., Massachusetts, 1966.

KREYSZIG, E., Advanced Engineering Mathematics, John Wiley & Sons, New York, ISBN-13: 978-0470458365, ISBN-10: 9780470458365, 2011.

LANG, S., Introduction to Linear Algebra, Addison-Wesley Publishing Co., Massachusetts, ISBN-13: 978-0387964126, ISBN-10: 0387964126, 1966.

PISKUNOV, N., Integral and Differential Calculus, Editorial MIR, Moscow, 1969.

STRANG, G., Linear Algebra and Its Applications, Academic Press, New York, 1976. Harcourt Brace Jovanovich (1980). Third Edition: Brooks/Cole (1988). Fourth Edition: Brooks/Cole/Cengage (2006).

3

MODELING AND ANALYSIS OF ELECTRICAL CIRCUITS WITH BLOCK DIAGRAMS

3.1 INTRODUCTION

In order to model energy conversion, power systems, and power electronic systems, the student has to understand at least the principles, foundations, and operation of the following: (i) passive components (inductors, capacitors, resistors, and transformers); (ii) semiconductors (diodes, transistors, and thyristors); (iii) analog controls; (iv) digital controls; (v) some power sources, such as voltage sources, current sources, batteries, rotating machine generators, photovoltaics, and fuel cell; and (vi) the basics of energy resources, such as fossil fuel-based thermodynamics, hydropower, wind, solar, and junction thermal devices. This chapter shows the use of block diagram-oriented simulation techniques for such electrical circuit topologies. The connecting blocks should be based on ordinary differential equations (ODEs) through a causal modeling approach. A causal model consists of a mathematical set of structural equations that can be represented by a flow diagram, where the information travels from one side of a particular function (block) to their output. However, the only way to make such information to travel backward is to invert mathematically the relationship. Block diagrams can usually represent ODEs for linear and nonlinear cases, and they became a very powerful computational paradigm for most of typical engineering problems.

It is important initially to understand how to model the following four elements: (i) an ideal switch, (ii) an ideal resistor, (iii) an ideal capacitor, and (iv) an ideal inductor.

Modeling Power Electronics and Interfacing Energy Conversion Systems, First Edition.
M. Godoy Simões and Felix A. Farret.
© 2017 John Wiley & Sons, Inc. Published 2017 by John Wiley & Sons, Inc.

The word "switch" has been used for mechanical elements, which open or close an electrical circuit. However, some switches are activated manually, others by an electrical coil (relay or contactors), or just electrically such as a transistor. An ideal switch can represent a semiconductor in totally ON or OFF states. Therefore, an ideal switch can replace diodes, transistors, and thyristors in understanding most of fundamental power electronic systems. For all switching elements, the following four states are considered:

1. Turned off
2. Turning on state
3. Turned on
4. Turning off state

In most cases, an infinite resistance can model the turned-off state (switch open), while the turned-on (switch closed) is represented by a zero (or small) resistance, and typically, those are the states we assume when any circuit is initially understood and modeled. The turning on/off states might be disregarded (considering an instantaneous transition from on to off and from off to on). However, they can be considered as a delay between the control signal and its real contact opening or closing, or by a linear or a nonlinear rise-time and fall-time transition, or carrying some tailing effects in the voltage or current, or maybe allowing an overshoot voltage because of some series leakage inductance. The complexity of such transition behavior can make the simulation very complex, sometimes requiring very specific physics-based models. This textbook does not encompass such physics-based modeling. Usually those physics-based models can be disregarded when understanding the main functions of power electronics, power systems, power quality, and renewable energy systems.

Resistors can be considered "ideal," that is, they simply obey the Ohm's law, where the voltage drop is the current across multiplied by the resistance, that is, $v = R \cdot i$. However, a nonideal behavior might be incorporated, such as assuming that the resistor can be dependent on (i) temperature, (ii) frequency (skin and proximity effect), (iii) inductive, (iv) capacitive, (v) aging, and (vi) nonlinear as a function of on their voltage and current or some other physical effect.

Capacitors are considered to be "ideal" when they follow a simple differential equation relationship, that is, $i = C \dfrac{dv}{dt}$, and their energy stored is calculated as $E_C = \dfrac{1}{2} C v^2$. Therefore, in steady state, there is no current flow, and when current is either positive or negative, the voltage will rise or decrease, and the instantaneous voltage is a measure of the energy stored. Capacitors may also have some complexities such as (i) temperature dependency, (ii) current-dependent losses ("equivalent serial resistance—ESR"), (iii) voltage-dependent losses (especially electrolytic capacitors), (iv) series inductance, and (v) double-layer capacitance.

Inductors are considered to be "ideal" when they follow the differential equation $v = L \dfrac{di}{dt}$, and their energy stored is calculated as $E_L = \dfrac{1}{2} L i^2$. Therefore, in steady state, there is no voltage drop, and when voltage is either positive or negative, the

TABLE 3.1   Analog Behavior of Passive Components

Resistance	Capacitance	Reluctance
$R = \rho \dfrac{l}{A}$	$C = \varepsilon \dfrac{A}{l}$	$\mathfrak{R} = \mu \dfrac{l}{A}$

current will rise or decrease, and the instantaneous current is a measure of the energy stored. Inductors may also have some complexities such as (i) losses (internal series resistance), (ii) frequency-dependent losses (because of skin effect, proximity effect, and magnetic core), (iii) nonlinearity because of magnetic core saturation effect, (iv) capacitive-like response, because the inductor turns are physically close; those near turns may respond for very high frequencies.

As discussed earlier, there is an analog behavior in these components, though the reader must understand that a very simple passive component can become very complicated to model, depending on several real-life effects to consider. The basic analog behavior is illustrated in Table 3.1 relating resistance, inductance, and reluctance. Therefore, causal modeling is very important in engineering analysis, and equation-oriented block diagrams can support a functional understanding of the system.

3.2   LABORATORY PROJECT: TRANSIENT RESPONSE STUDY AND LAPLACE TRANSFORM-BASED ANALYSIS BLOCK DIAGRAM SIMULATION

Figure 3.1 shows the same electrical circuit used in Figure 2.1, where frequency-dependent passive components are represented by their equivalent Laplace impedance, that is, sL for inductances and $\dfrac{1}{sC}$ for capacitances, plus external sources are assumed to be generally $V_0(s)$ and $I_x(s)$. The use of the Laplace transform-based approach for circuit analysis allows the derivation of any circuit analysis using algebraic equations. Then the analysis can aid the drawing of a block diagram for Simulink simulation or allow frequency response analysis by letting $s = j\omega$.

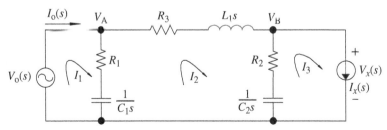

$R_1 = 10\ \Omega$, $R_2 = 1.5\ \Omega$, $R_3 = 5\ \Omega$, $C_1 = 200\ \mu F$, $C_2 = 400\ \mu F$,   and $L_1 = 750\ mH$

FIGURE 3.1   Laplace transform-based circuit analysis.

The circuit depicted in Figure 3.1 has three mesh currents solved by the following three equations:

- Mesh #1

$$-V_0(s) + R_1\left[I_1(s) - I_2(s)\right] + \frac{1}{sC_1}\left[I_1(s) - I_2(s)\right] = 0 \tag{3.1}$$

- Mesh #2

$$-\frac{1}{sC_1}\left[I_1(s) - I_2(s)\right] - R_1\left[I_1(s) - I_2(s)\right] + R_3 I_2(s) + sL_1 I_2(s)$$
$$+R_2\left[I_2(s) - I_3(s)\right] + \frac{1}{sC_2}\left[I_2(s) - I_3(s)\right] = 0 \tag{3.2}$$

- Mesh #3

$$I_3(s) = I_x(s) \tag{3.3}$$

The matrix form of the mesh equations is

$$\begin{bmatrix} \left(R_1 + \dfrac{1}{sC_1}\right) & -\left(R_1 + \dfrac{1}{sC_1}\right) & 0 \\ -\left(R_1 + \dfrac{1}{sC_1}\right) & \left(R_1 + R_2 + R_3 + \dfrac{1}{sC_1} + \dfrac{1}{sC_2} + sL_1\right) & -\left(R_2 + \dfrac{1}{sC_2}\right) \\ 0 & 0 & 1 \end{bmatrix}\begin{bmatrix} I_1(s) \\ I_2(s) \\ I_3(s) \end{bmatrix} = \begin{bmatrix} V_0(s) \\ 0 \\ I_x(s) \end{bmatrix} = $$

$$Z\begin{bmatrix} I_1(s) \\ I_2(s) \\ I_3(s) \end{bmatrix} = \begin{bmatrix} V_0(s) \\ 0 \\ I_x(s) \end{bmatrix} \tag{3.4}$$

In Chapter 2, the objective was to arrive to a linear matrix formulation $ZI = V$ and allow calculation in steady state using phasor analysis. However, the goal of the current procedure is to obtain a state–space formulation $\dot{x} = Ax + Bu$. Defining the state variables as the current through the inductor and the voltage across the capacitors, we will need first-order equations with $\dfrac{di_{L_1}}{dt}$, $\dfrac{dv_{C_1}}{dt}$, and $\dfrac{dv_{C_2}}{dt}$ isolated on the left-hand side of their equations. Therefore, the previous equations must be expanded by their common denominators, and after algebraic manipulation, we should isolate the state variables (capacitor voltages and inductor current), calculated from their system mesh currents $I_1(s)$, $I_2(s)$, and $I_3(s)$, keeping them on the left side of a system equation, together with the Laplace operator multiplying them. That would make their time-domain derivatives equivalent to the states–space formulation, that is, the objective is to obtain

$$\begin{bmatrix} sV_{C_1}(s) \\ sV_{C_2}(s) \\ sI_{L_1}(s) \end{bmatrix} = A_{3x3}(s) \begin{bmatrix} V_{C_1}(s) \\ V_{C_2}(s) \\ I_{L_1}(s) \end{bmatrix} + B_{3x2}(s) \begin{bmatrix} V_0(s) \\ I_x(s) \end{bmatrix} \tag{3.5}$$

where $V_0(s)$ and $I_x(s)$ are the Laplace transform of the electrical power sources delivering energy to the circuit.

In order to convert the mesh equations into state–space equations, it is necessary the relationship of the appropriate variables. For this circuit, we have

$$V_{C_1}(s) = \frac{1}{sC_1} \left[I_1(s) - I_2(s) \right] \tag{3.6}$$

$$V_{C_2}(s) = \frac{1}{sC_2} \left[I_2(s) - I_3(s) \right] \tag{3.7}$$

$$I_{L_1}(s) = sL_1 I_2(s) \tag{3.8}$$

$$\begin{bmatrix} V_{C_1}(s) \\ V_{C_2}(s) \\ I_{L_1}(s) \end{bmatrix} = \begin{bmatrix} \frac{1}{sC_1} & -\frac{1}{sC_1} & 0 \\ 0 & \frac{1}{sC_2} & -\frac{1}{sC_2} \\ 0 & sL_1 & 0 \end{bmatrix} \begin{bmatrix} I_1(s) \\ I_2(s) \\ I_3(s) \end{bmatrix} = T \begin{bmatrix} I_1(s) \\ I_2(s) \\ I_3(s) \end{bmatrix} \tag{3.9}$$

and

$$\begin{bmatrix} I_1(s) \\ I_2(s) \\ I_3(s) \end{bmatrix} = T^{-1} \begin{bmatrix} V_{C_1}(s) \\ V_{C_2}(s) \\ I_{L_1}(s) \end{bmatrix} = \begin{bmatrix} \frac{1}{sC_1} & -\frac{1}{sC_1} & 0 \\ 0 & \frac{1}{sC_2} & -\frac{1}{sC_2} \\ 0 & sL_1 & 0 \end{bmatrix}^{-1} \begin{bmatrix} V_{C_1}(s) \\ V_{C_2}(s) \\ I_{L_1}(s) \end{bmatrix} \tag{3.10}$$

Therefore, substituting Equation 3.10 into 3.4 results to

$$
\mathbf{Z}\begin{bmatrix} I_1(s) \\ I_2(s) \\ I_3(s) \end{bmatrix} = \mathbf{ZT}^{-1} \begin{bmatrix} V_{C_1}(s) \\ V_{C_2}(s) \\ I_{L_1}(s) \end{bmatrix}
$$

$$
= \begin{bmatrix} \left(R_1 + \dfrac{1}{sC_1}\right) & -\left(R_1 + \dfrac{1}{sC_1}\right) & 0 \\ -\left(R_1 + \dfrac{1}{sC_1}\right) & \left(R_1 + R_2 + R_3 + \dfrac{1}{sC_1} + \dfrac{1}{j\omega C_2} + sL_1\right) & -\left(R_2 + \dfrac{1}{sC_2}\right) \\ 0 & 0 & 1 \end{bmatrix}
$$

$$
\times \begin{bmatrix} \dfrac{1}{sC_1} & -\dfrac{1}{sC_1} & 0 \\ 0 & \dfrac{1}{sC_2} & -\dfrac{1}{sC_2} \\ 0 & sL_1 & 0 \end{bmatrix}^{-1} \times \begin{bmatrix} V_{C_1}(s) \\ V_{C_2}(s) \\ I_{L_1}(s) \end{bmatrix}
$$

(3.11)

Although matrix Equation 3.11 has now the correct state–space variables, it still needs algebraic work to change into a state–space formulation. For small systems (two or three nodes or meshes), it is possible to expand each equation (each row) and work to have $sV_{C_1}(s)$, $sV_{C_2}(s)$, and $sI_{L_1}(s)$ on the left-hand side. For larger systems, we can use symbolic software, such as Mathematica or Maple, in order to isolate $sV_{C_1}(s)$, $sV_{C_2}(s)$, and $sI_{L_1}(s)$. In general, the student should decide if he or she wants to pursue either (i) a mesh/nodal analysis or (ii) a state–space analysis in order to analyze an electrical circuit and avoid such algebraic hand or symbolic manipulation. For example, the circuit portrayed in Figure 3.1 has an easier evaluation if a combination of KVL and KCL are used to provide the voltages and current actually required for state–space analysis. So, the analysis must be centered on the derivatives of capacitor voltages, inductor currents, and their relationship to their own state variables and power sources delivering the electrical energy to the circuit:

$$
\frac{dv_{c_1}}{dt} = \frac{1}{C_1} \cdot \frac{v_0 - v_{c_1}}{R_1} \tag{3.12}
$$

$$
\frac{dv_{c_2}}{dt} = \frac{i_{L_1} - i_x}{C_2} \tag{3.13}
$$

$$
\frac{di_{L_1}}{dt} = \frac{1}{L_1}\left(v_0 - i_{L_1}R_2 - \left(i_{L_1} - i_x\right)R_3 - v_{c_2}\right) \tag{3.14}
$$

The state–space matrix equation is

$$\frac{d}{dt}\begin{bmatrix} v_{c_1} \\ v_{c_2} \\ i_{L_1} \end{bmatrix} = \begin{bmatrix} \dfrac{-1}{C_1 R_1} & 0 & 0 \\ 0 & 0 & \dfrac{1}{C_2} \\ 0 & \dfrac{-1}{L_1} & \dfrac{-(R_2+R_3)}{L_1} \end{bmatrix}\begin{bmatrix} v_{c_1} \\ v_{c_2} \\ i_{L_1} \end{bmatrix} + \begin{bmatrix} \dfrac{1}{C_1 R_1} & 0 \\ 0 & \dfrac{-1}{C_2} \\ \dfrac{1}{L_1} & \dfrac{R_2}{L_1} \end{bmatrix}\begin{bmatrix} v_o \\ i_x \end{bmatrix} \quad (3.15)$$

The system is augmented for the calculation of the following output variables:

$$\begin{bmatrix} i_{c_1} \\ i_{c_2} \\ i_o \\ v_B \\ v_{L_1} \\ v_{R_1} \\ v_{R_3} \end{bmatrix} = \begin{bmatrix} \dfrac{-1}{R_1} & 0 & 0 \\ 0 & 0 & 1 \\ \dfrac{-1}{R_1} & 0 & 1 \\ 0 & 1 & R_2 \\ 0 & -1 & -(R_2+R_3) \\ -1 & 0 & 0 \\ 0 & 0 & R_3 \end{bmatrix}\begin{bmatrix} v_{c_1} \\ v_{c_2} \\ i_{L_1} \end{bmatrix} + \begin{bmatrix} \dfrac{1}{R_1} & 0 \\ 0 & -1 \\ \dfrac{1}{R_1} & 0 \\ \dfrac{1}{R_1} & -R_2 \\ 1 & R_2 \\ 1 & 0 \\ 0 & 0 \end{bmatrix}\begin{bmatrix} v_o \\ i_x \end{bmatrix} \quad (3.16)$$

A block diagram representing the circuit is depicted on Figure 3.2. The Simulink implementation with scope measurements for output variables and save to workspace data storage is shown on Figure 3.3.

```
%%Parameters
f = 60; w=2*pi*f;
v0 = 169.7; v0Angle = 0.25;
ix = 1.5; ixAngle = -(pi/2)-0.15;
R1 = 10;R2 = 1.5; R3 = 5; C1 = 200e-6; C2 = 400e-6; L=750e-3;
X1 = w*L;Xc1=1/(w*C1);Xc2=1/(w*C2);
%State Space Model
A = [-1/(C1*R1)              0           0
     0                       0           1/C2
     0                      -1/L        -(R2+R3)/L];
B = [1/(C1*R1)              0
     0                     -1/C2
     1/L                    R2/L];
C = [-1/R1          0        0
      0             0        1
     -1/R1          0        1
      0             1        R2
      0            -1       -(R2+R3)
     -1             0        0
      0             0        R3];
```

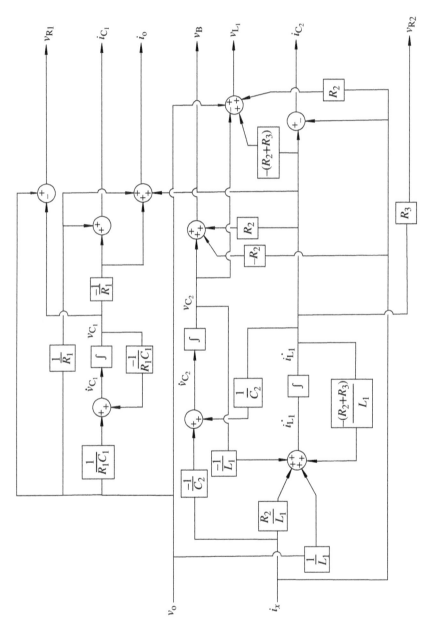

FIGURE 3.2 Block diagram for electrical circuit of Figure 3.1.

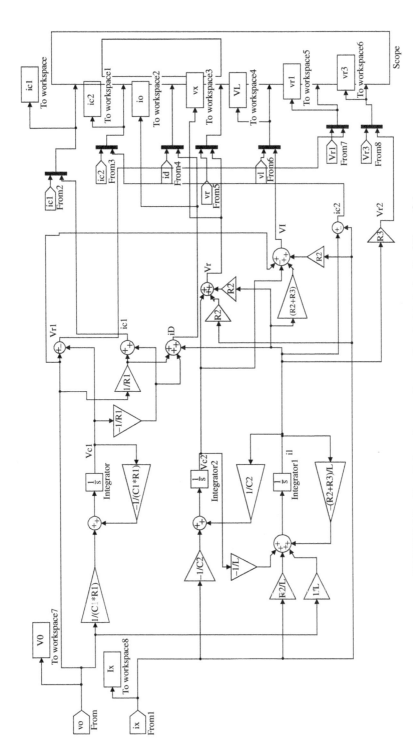

FIGURE 3.3 Simulink workspace diagram for electrical circuit of Figure 3.1.

```
D = [1/R1  0
     0                    -1
     1/R1                  0
     0                    -R2
     1                    R2
     1                     0
     0                     0 ];
%%end of Matlab script  -------------------------------
```

3.3 COMPARISON WITH PHASOR-BASED STEADY-STATE ANALYSIS

The block diagram simulation can be compared with the phasor analysis of Chapter 2 by assuming that $v_0(t) = 169.7\cos(2\pi 60t + 0.25)$ (V) and $i_x(t) = 1.5\sin(2\pi 60t - 0.15)$ (A). The current source can also be easily flipped in polarity as $i_x(t) = -1.5\sin(2\pi 60t - 0.15)$. All the MATLAB-based linear calculations using phasor analysis can be checked against the time-domain solutions of this block diagram-based simulation. Although the workspace diagram of Figure 3.3 does not contain the power factor computation, it can be easily incorporated in the simulation study. Figures 3.4, 3.5, 3.6 and 3.7 show voltages and currents waveforms simulated with this block diagram simulation methodology.

A transient response simulation such as the one done via a Simulink block diagram has several advantages:

1. Any parameter can be easily changed, made variable, time dependent, temperature dependent, nonlinear, and so on.
2. The input sources, in this case $v_0(t)$ and $i_x(t)$ can be selected from a large and diverse possibilities in the Simulink library, and several types of transient responses, sharp variation, and addition of high-frequency terms can be incorporated in the simulation.
3. The Simulink and MATLAB environments are seamless interconnected and interoperating. Therefore, parameters and lookup tables can be read from the MATLAB workspace, data can be saved and retrieved, MATLAB functions can add functionalities to the Simulink-based analysis, compiled code can be built in functions, and digital control can be incorporated in the loop.
4. Several other toolboxes available in MATLAB can be used in the simulation study such as control systems, signal processing, power systems, fuzzy logic, artificial neural networks, optimization, and so on and so forth.
5. With this Simulink-based study, it becomes very simple to define the equivalent Thévenin, as suggested in Chapter 2, with some case studies, by considering the voltage across the terminals to find equivalent Thévenin model the following conditions: (i) open-circuit response and (ii) short-circuit response.

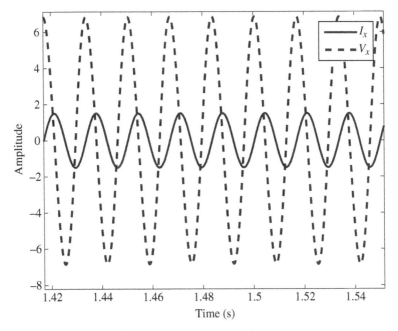

FIGURE 3.4   Voltage and current of current source.

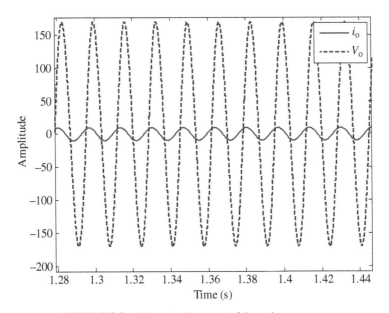

FIGURE 3.5   Voltage and current of the voltage source.

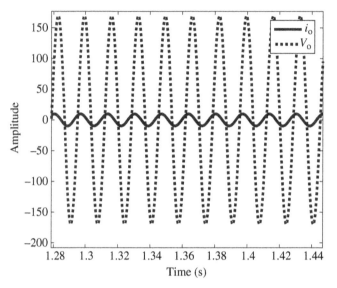

FIGURE 3.6   Voltage and current of the voltage source with changing direction of the current source.

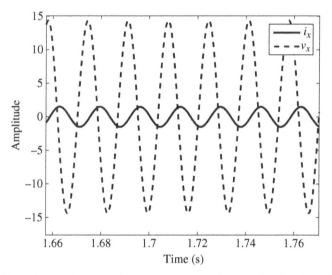

FIGURE 3.7   Voltage and current of the current source with changing direction of the current source.

3.4   FINDING THE EQUIVALENT THÈVENIN

Suppose the just-discussed simulation study can be used to find the equivalent Thévenin for R_2. The procedure, always valid for linear systems, even if they have dependent voltage or current sources, is to measure the open-circuit voltage, in this case across the terminals of R_3 when it is removed, after finding the correct time-domain representation and considering a Thévenin equivalent voltage $v_{TH}(t)$. Then it is necessary to measure the short-circuit current across same terminals, $i_{SC}(t)$, and eventually calculate the equivalent Thévenin impedance by ratio of two phasor quantities derived from the time-domain observation of $v_{TH}(t)$ and $i_{SC}(t)$:

$$\frac{\bar{V}_{TH} = |\bar{V}_{TH}|\angle\varnothing_{TH}}{\bar{I}_{SC} = |\bar{I}_{SC}|\angle\varnothing_{SC}} = Z_{TH} = \left(R_{eq} + jX_{eq}\right) \tag{3.17}$$

Using the Simulink model,

$$v_{oc} = 159.669\angle75.03° \qquad\qquad i_{sc} = 0.578\angle-74.7°$$

The Thévenin impedance is the ratio of the Thévenin voltage (also open-circuit voltage) divided by the short-circuit current:

$$Z_{TH} = \frac{v_{oc}}{i_{sc}} = 276.244\angle149.73° \tag{3.18}$$

$$V = \frac{R_3}{Z_{TH} + R_3} \times v_{oc} = 2.93\angle-74.17° \tag{3.19}$$

The dissipated power is

$$P = \frac{2.93^2}{5} = 1.72\,\text{W} \tag{3.20}$$

In Simulink, the resistance R_3 can be made very high (when compared to the other resistances, in order to simulate open-circuit conditions. Then, $\bar{V}_{TH}$ is measured as the open-circuit voltage ($\bar{V}_{TH} = 159.669\angle75.03°$). After that, R_3 can be made very low in order to simulate short-circuit current conditions, and the short-circuit current is found as $\bar{I}_{SC} = 0.578\angle-74.72°$. From these measurements, a simple calculation allows the calculation of the equivalent Thévenin impedance and the dissipated power across R_3. Of course, an equivalent Thévenin model shown in Figure 3.8 can be useful for many purposes.

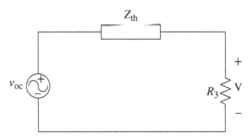

FIGURE 3.8 Equivalent Thévenin across R_3 was found via a simulation-based study of the system block diagram in Simulink.

3.5   SUGGESTED PROBLEMS

1. Build simple block diagram models (in Simulink) and compare with their time-domain solutions in mathematical equations (in MATLAB) for the following cases:

    i. A series RL connected with a voltage source with a switch to turn on and turn off.

    ii. A series RC connected with a current source; the current source is either in short circuit or instantaneously connected to the RC with a switch.

    iii. A parallel RLC with a current source; the current source is either in short circuit or instantaneously connected to the RLC with a switch.

    iv. A series RLC with a voltage source with a switch to turn on and turn off.

2. The circuit of Figure 3.9 is a series resonant circuit (also called as harmonic trap filter).

    i. Implement a simulation study for this circuit with a block diagram in Simulink.

    ii. Vary the voltage excitation angular frequency ω from 0 to 2000 rad/s, save one or two cycles of $i(t)$ after reaching steady state, make a MATLAB analysis of all those responses, and compare with the expected theoretical conclusions for such resonant circuit.

    iii. Assume that this circuit has no external voltage source; it is only a series connected RLC. Study the response with initial conditions v_c (0_) and $i_L(0\_)$, and compare with expected theoretical analysis.

    iv. Calculate the time-domain response of a resonant series RLC with initial conditions, implement the state equation in MATLAB, and compare with the Simulink-based response.

    v. Write a design procedure (in MATLAB) for this filter that allows a particularly harmonic to be trapped by the filter. Calculate the power loss of such filter when inserted in parallel with a circuit dependent on their parameters.

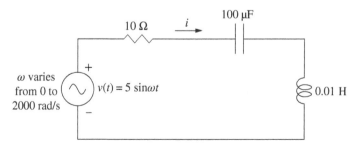

FIGURE 3.9   Series resonant (trap filter) circuit.

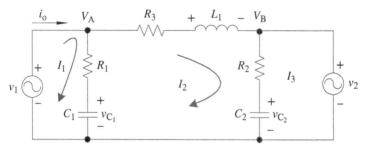

FIGURE 3.10 Model of a short transmission line with two synchronous generators.

3. The circuit in Figure 3.10 represents the model of a transmission line with two synchronous generators in steady-state, $\overline{V_1} = |\overline{V_1}| \angle \varnothing_1$ and $\overline{V_2} = |\overline{V_2}| \angle \varnothing_2$. Implement a simulation study for this circuit as a block diagram in Simulink. The active power will transfer from one source to another depending on the phase shift, and the reactive power will circulate from one source to another depending on the voltage amplitude. Study how this system operates using basic power systems fundamentals, observing how it works in Simulink and comparing their steady-state operation in MATLAB.

4. Design advanced passive component models, with the following assumptions:

   i. Resistors can be considered "ideal," that is, they simply obey the Ohm's law, where the voltage drop is the current across multiplied by the resistance, that is, $v = R \cdot i$. However, nonideal behavior might be incorporated such as assuming the resistors to be dependent on (i) temperature, (ii) frequency (skin effect), (iii) inductive, and (iv) nonlinearity depending on their voltage and current.

   ii. Capacitors are considered "ideal" when they follow a simple differential equation relationship, that is, $i = C \dfrac{dv}{dt}$, and their energy stored is calculated as $E_C = \dfrac{1}{2} C v^2$. Therefore, in steady state, there is no current flow, and when current is either positive or negative, the voltage will rise or decrease, and the instantaneous voltage is a measure of the energy stored. Capacitors may also have some complexities such as (i) temperature dependent, (ii) current-dependent losses ("equivalent serial resistance—ESR"), (iii) voltage-dependent losses (especially electrolytic capacitors), (iv) series inductance, and (v) double-layer capacitance.

   iii. Inductors are considered to be "ideal" when they follow the differential equation $v = L \dfrac{di}{dt}$, and their energy stored is calculated as $E_L = \dfrac{1}{2} L i^2$. Therefore, in steady state, there is no voltage drop, and when voltage is either positive or negative, the current will rise or decrease, and the instantaneous current is a measure of the energy stored. Inductors may also have some complexities such as (i) losses (internal series resistance),

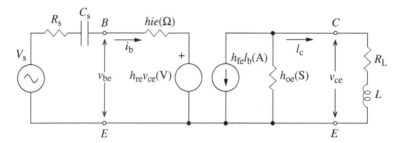

FIGURE 3.11   Hybrid model of a bipolar junction transistor (BJT).

(ii) frequency-dependent losses (because of skin effect, proximity effect, and magnetic core), (iii) nonlinearity because of magnetic core saturation effect, and (iv) capacitive-like response, because the inductor turns are physically close and may respond for very high frequencies.

5. The circuit represented in Figure 3.11 is the hybrid model of a bipolar junction transistor. Find out an expression for the load voltage as a function of the base current and establish the analysis and simulation for this circuit including the load. After that, plot in MATLAB the linear output characteristics relating the collector current I_c with the collector–emitter voltage V_{ce} as a function of the base current I_b in five levels. Finally, establish the load line with a quiescent point on the middle of it and verify what is the collector current and V_{ce} for this point.

h_{ie}	$510\,\Omega$
h_{re}	2.5×10^{-4}
h_{fe}	50
h_{oe}	$25\,\mu\text{A/V}$
C	$100\,\mu\text{F}$
L	$0.2\,\text{mH}$
R_L	$470\,\Omega$
V_s	$3.5\sin(\omega t)$
f_s	$1000\,\text{Hz}$

FURTHER READING

CLOSE, C., FREDERICK, D.K. and NEWELL, J.C., Modeling and Analysis of Dynamic Systems, John Wiley & Sons, New York, ISBN-13: 978-0471394426, 2002.

DESOER, C.A. and KUH, E.S., Basic Circuit Theory, McGraw-Hill Book Co., New York, 2009.

KARRIS, S., Circuit Analysis I, with MATLAB Computing and Simulink/SimPowerSystems Modeling, Orchard Publications, Fremont, ISBN-13: 978-1934404171, ISBN-10: 1934404179, 2009.

KARRIS, S., Circuit Analysis II, with MATLAB Computing and Simulink/SimPowerSystems Modeling, Orchard Publications, Fremont, ISBN-13: 978-1934404195, ISBN-10: 1934404195, 2009.

KLEE, H. and ALLEN, R., Simulation of Dynamic Systems with MATLAB and Simulink, 2nd edition, CRC Press, Boca Raton, ISBN-13: 978-1439836736, ISBN-10: 1439836736, 2011.

KREIDER, D.L., KULLER, R.G., OSTBERG, D.R., PERKINS, F.W. and LOOMIS, L.H., An Introduction to Linear Analysis, Addison-Wesley Series in Mathematics, Boston, ISBN-10: 020103946X, ISBN-13: 978-0201039467, 1966.

KREYSZIG, E., Advanced Engineering Mathematics, John Wiley & Sons, New York, ISBN-13: 978-0470458365, ISBN-10: 9780470458365, 2011.

MALLEY, J.O., Theory and Problems of Basic Circuit Analysis, Schaum's Outline Series, McGraw-Hill, Inc., New York, 1982.

MOHAN, N., Advanced Electric Drives: Analysis, Control, and Modeling Using MATLAB/ Simulink, John Wiley & Sons, Inc., Hoboken, ISBN: 978-1-118-48548-4, 2014.

PATON, B.E., Sensors, Transducers & LabVIEW, Prentice Hall PTR, New Jersey, ISBN-10: 0130811556, ISBN-13: 978-0130811554, ISBN-10: 0130811556, 1998.

WOODS, R.L. and LAWRENCE, K.L., Modeling and Simulation of Dynamic Systems, Prentice Hall, Upper Saddle River, ISBN: 0-13-337379-7, 1997.

4

POWER ELECTRONICS: ELECTRICAL CIRCUIT-ORIENTED SIMULATION

4.1 INTRODUCTION

In a circuit-oriented simulation environment, there are usually three levels of modeling: (1) component level, whose purpose is to consider switching transients, parasitic effects, where detailed physical switching models are required, such as diode reverse recovery effects, losses, stray capacitances, and leakage inductances; (2) circuit level, where the goal is to study the converter functionality, outer/inner loop design, where models should consider the behavioral switch models assuming instantaneous switching; and finally, (3) system level, where the interaction of system modules and smart functionalities of the outer controllers can be explored, aiming further user-interfacing control. For system-level simulation, only a detailed transfer function representation of converters would be required in modeling. Most circuit simulators are made only for level 1 (such as PSPICE-based simulators) or for level 3 (block diagram, Simulink or similar simulators). For level 2 (most of projects in this book), PSIM is used as a good circuit simulator that encompasses capabilities of interfacing with built-in toolboxes or C-program for level 3. The Power Systems Toolbox of MATLAB is a level 3 simulator with some functionalities to operate at level 2, and full-fledged integration with MATLAB and Simulink allows even more level 3 capabilities.

A system-level simulation can merge the interaction of different modules in a system, for example, a converter, a controller, source, and load. Another example

Modeling Power Electronics and Interfacing Energy Conversion Systems, First Edition.
M. Godoy Simões and Felix A. Farret.
© 2017 John Wiley & Sons, Inc. Published 2017 by John Wiley & Sons, Inc.

would be a motor drive with mechanical load, electromagnetics, power electronics, digital control, and supervisory control. At this level, the low-level switching behavior of a power converter has no major concern, and a converter, or a machine, could be represented by a transfer function or by a black box model where only parameters are used. The Power Systems Toolbox of MATLAB has several complex power converters, machines, or transformers, available with black box models, and several filters, and controllers, could be defined by their transfer function or by their parameters (such as proportional, integral, and derivative gains). Multi-domain simulators will certainly aggregate a very high-level system approach. Some power systems-oriented simulators (such as PSCAD, DlgSILENT, EMTP-RV, PSS/E) have a system-level approach.

A very appropriate level of circuit-level modeling must be chosen to incorporate the functionality and large signal behavior of switching converters, fast inner control loops, and fundamental power electronic switch operation (such as incorporating voltage drop plus rise and fall time characteristics). Those would use semiconductors with simplified behavioral or quasi-ideal models. The PSIM circuit simulator approaches the idealized switching of semiconductors instead of detailed physical models with a good degree of analog and digital circuits required for designing circuits. It has also flexibility in designing a real-time control by incorporating C-language and allowing the export of the controller's code for DSP hardware-based implementation.

The component-level modeling studies device voltage and current transients, turn-on and turn-off losses, and impact of switching on overvoltage on leakage inductances. Such behavior that occurs in the order of nanoseconds influences the microelectronics and integrated circuit manufacturing or component-level design. However, those characteristics are not the first concern of undergraduate students or even professionals who wish to understand how a circuit or a system operates. The first concern is how the topology and basic controls make the electrical and electronic circuit operational. On the other hand, for component-level model, device models and precise circuit stray parameters are required in addition to electromagnetics high frequency-based design studies and electromagnetic compatibility studies. A SPICE-oriented simulator would be the framework for such component-level modeling, but there are other simulators concerned with high-frequency and microelectronics studies, (such as SABER simulation platform). Component-level modeling is not the main approach in this book; instead, the authors decided to make a firm foundation for students and readers to have a methodology in understanding how the power circuit and associate analog or digital controls will make the overall system to properly work. The current chapter will concentrate on circuit-level approach, and in several other chapters, the system-level will be the focus.

In this textbook, we want to develop modeling and computational skills that allow the reader to study how a power electronic subsystem interacts with the connected systems using a careful reduction in the system complexity by some of the following considerations:

a) Using one or a few equivalent models or approximated devices in series or in a parallel combination.

b) Representing power electronic nonlinear loads with their similar characteristics, with a good precision and timing representation.

c) Understanding what is the range from the simplest possible model to the most advanced one and considering how those extremes of complexity will affect the system analysis.

d) Making the representation of any power electronic subsystem by their equivalent voltage or current sources (linear or nonlinear), or using a transfer function or maybe a black box I/O model constructed out of data analysis, as long as this is acceptable and makes sense for the particular study.

e) Representing only a front end of the full subsystem interconnected to a particular point of interest, when the major concern is utility interfacing or microgrid operation.

f) Including system dynamic and controls only when necessary and making decisions when inner control loops are not relevant for a system-level application.

g) Eventually, understanding how to use a modular approach, initially assuming casual systems, and then applying techniques that support acausal system modeling. When blocks can be reused, when code can be recoded for real-time applications, and when different domains, such as thermal, mechanical, and electrical, can be recycled in further applications, the designer develops a strong comprehension, and he or she trusts on previous good modeling efforts and can work toward larger and sparser model developments approaches.

A simple circuit that can be studied in a circuit simulator is a half-wave rectifier, because it can be used for cross-examination of exact mathematical calculations (by hand) with their simulations in a circuit simulator (such as PSIM) as well as their numerical computation using a platform such as MATLAB.

Suppose the half-wave rectifier indicated in Figure 4.1 is implemented in PSIM, and the main objective is to observe all the average and effective values of voltages and currents, as well as average power delivered to the load and the apparent power taken from the sinusoidal source.

In this book, we discussed some ideas of either using DAE or ODE (Chapter 1) to represent acausal and causal systems. Block diagrams may have algebraic loops (variables that depend on themselves) indicating an ill-modeling or maybe

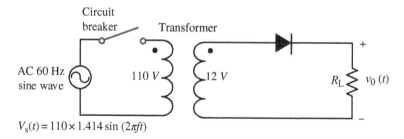

$$V_s(t) = 110 \times 1.414 \sin(2\pi ft)$$

FIGURE 4.1 Half-wave rectifier connected to an AC power source.

the negligence of time response, but that may also be an acausal system. Acausality, in signal processing, is when the production of an output signal which is processed from another input signal, depends on the recording of input values either forward or backward in the sampled time series, from a particular and arbitrary predefined time denoted as "present" time. In signal processing, such "present" time input, as well as the "future" time input values, has been recorded at some time in the past, but conceptually it can be called the "present" or "future" input values in this acausal process. This type of processing cannot be done in real-time, as future input values are not yet known, but is done after the input signal has been recorded and is postprocessed. Optimal control with the solution of the Riccati equation is a good example of an acausal system.

The time-domain voltages and currents require some specific values that define their features; some particular ones are the calculation of average and the effective values, which can indicate an equivalent direct current (DC) parameter or an equivalent alternating current (AC) parameter for any instantaneous quantity. The instantaneous multiplication of voltage and current will usually result in a very complicated instantaneous power quantity. Therefore, the average of power could be calculated, that is, how much power becomes either heat or motion. However, there is circulation of currents in any system that does not reach the load, and they contribute for increasing the apparent power taken from any source. Therefore, from the utility point of view, or the provider of electrical power, it is very important for them to know the effective value of the current, so the calculation of the apparent power as the multiplication of effective value of voltage times the effective value of current is relevant, because such apparent power is proportional to the utility costs and investments in their infrastructure and maintenance. The ratio of average power (also called real power) to the apparent power is defined as power factor (PF), and it is a very important figure of merit to characterize the quality of the power delivery from source to the load.

A general definition for the average and effective values for a time-domain variable $x(t)$ are

$$\langle x \rangle = X_{\text{AVG}} = \frac{\int_0^T x(t)\,dt}{T} \tag{4.1}$$

$$\langle x^2 \rangle = \frac{\int_0^T x^2(t)\,dt}{T} \rightarrow X_{\text{RMS}} = \sqrt{\frac{\int_0^T x^2(t)\,dt}{T}} \tag{4.2}$$

Very often the average value is called by X_{DC} or X_{AV}, and the effective value is called by X_{AC} or X_{RMS}. These parameters can only be calculated for periodic signals. Therefore, a time-changing signal, which might be variable in amplitude or in frequency, or a very nonlinear signal, might not have an evaluation of average and effective values as defined by Equations 4.1 and 4.2. Of course, when a closed algebraic formulation for $x(t)$ is known, we can easily calculate (4.1) and (4.2) by hand. In addition, the literature shows the results for several typical waveforms, such as sinusoid, square wave, triangular, sawtooth, and pulsed waveforms in textbooks.

For more complicated periodical waveforms, probably a symbolic software (such as Mathematica or Maple) can be useful in solving the integrals. The sinusoidal case is very well considered in power systems, that is, for $y(t) = A_{peak} \cos(\omega t + \phi)$ or similarly when $y(t) = A_{peak} \sin(\omega t + \phi)$, the solution for Equations 4.1 and 4.2 will give the results as $Y_{AVG} = 0$ and $Y_{RMS} = \dfrac{A_{peak}}{\sqrt{2}}$, that is, for pure sinusoidal waveforms.

For sampled waveforms, that is, data stored and available for processing, it is necessary to implement a discrete calculation of average and effective values. Therefore, the integration can be implemented by a trapezoidal rule, also called as midpoint rule. A possible calculation requires N samples acquired at instants $t_0, t_1, t_2, \dots t_{n-1}$, equidistant and distributed during one period T. The average is calculated by a simple sum over a period and division by the number of points:

$$Y_{AVG(t_{i+1})} = \frac{1}{N}\sum_{k=0}^{N-1} y\left(t_{i-k}\right) \tag{4.3}$$

The root-mean-square (RMS) value at the time t_i, is computed depending on the last N sampled values of the analog signal $y(t)$ such as

$$Y_{RMS(t_i)} = \sqrt{\frac{1}{N}\sum_{k=1}^{N-1} y^2\left(t_{i-k}\right)} \tag{4.4}$$

A recursive way to calculate a new effective value at t_{i+1} can be made by just incorporating the newest measurement $y(t_{i+1})$ and discarding the oldest measurement at $y(t_{i-N+1})$ as

$$\left(Y_{RMS(t_{i+1})}\right)^2 = \left(Y_{RMS}, (t_i)\right)^2 + \frac{1}{N}\left[y^2\left(t_{i+1}\right) - y^2\left(t_{i-N+1}\right)\right] \tag{4.5}$$

The main advantages of such recursive RMS method are reduced computational load and the correction of some errors of the analog-to-digital conversion. Such method to find a direct RMS value from previous RMS computation, over the last period, does not have a high precision, but it is good for an instantaneous estimation with precision from 4 to 6%, depending if the signal $y(t)$ does not change too fast.

MATLAB has already built-in functions that calculate those parameters. The MATLAB functions are called **mean** (for average value) and **RMS** (for effective value). Mean can be used with arrays, and it has commands to average either the rows or the columns, for example,

$$A = \begin{bmatrix} 1 & 2 & 6 \\ 4 & -7 & 0 \end{bmatrix}$$

```
A=[1,2,6;4,-7,0];
B=mean(A,1);
C=mean(A,2);
```

B =

   2.5000 −2.5000 3.0000

C =

   3

   −1

The function Y = rms(X) returns the RMS level of the input, X. If X is a row or column vector, Y is a real-valued scalar. For matrices, Y contains the RMS levels computed along the first nonsingleton dimension. For example, if X is an N-by-M matrix with N > 1, Y is a 1-by-M row vector containing the RMS levels of the columns of X.

Y = rms(X, DIM) computes the RMS level of X along the dimension, DIM.

An important function in MATLAB used to calculate a numerical integration is trapz. The syntax is

Z = trapz(Y)

Z = trapz(X,Y)

Z = trapz(…,dim)

Z = trapz(Y) computes an approximation of the integral of Y via the trapezoidal method (with unit spacing).

To compute the integral for spacing other than one, multiply Z by the spacing increment.

If Y is a vector, trapz(Y) is the integral of Y.

If Y is a matrix, trapz(Y) is a row vector with the integral over each column.

Z = trapz(X,Y) computes the integral of Y with respect to X using trapezoidal integration.

One example is to find the numerical computation for $Z = \int_{0}^{\pi} \sin(x)dx$, whose exact value is Z = 2. In order to approximate this integration numerically, one can initially make the X and Y vectors as

X = 0:pi/100:pi;

Y = sin(x);

Two possible ways to use the trapezoidal rule of integration can be

Z = trapz(X,Y)

or:

Z = pi/100*trapz(Y)

They give the following numerical computation for the definite integral as

Z =1.9998

4.2 CASE STUDY: HALF-WAVE RECTIFIER

The electrical circuit of a half-wave rectifier is depicted in Figure 4.1, and its PSIM simulation is indicated in Figure 4.2.

The objective of this simulation is to compare analytical solutions with PSIM-based studies with a numerical calculation in MATLAB of several RMS and average values of voltages, currents, active, reactive, and apparent powers. The PSIM simulation has voltage and current measurements indicated in Figures 4.3 and 4.4. In order to export data from PSIM to a MATLAB file, we can go to *edit* and choose *view data points*. Figure 4.4 shows the input sinusoidal source voltage and the corresponding half-wave rectifier voltage. Figure 4.5 shows the load voltage across the resistor and how PSIM can be used for finding the average of that waveform. Figure 4.6 is similarly the load voltage across the resistor with the PSIM calculation of the RMS value for such waveform. Therefore, by saving this data and import it to MATLAB, we are able to plot waveforms in MATLAB and make other calculations.

The reader can record the load-rectified voltage for one or two full cycles and calculate in PSIM their average and effective voltages. Their ratio can be analytically evaluated as Equation 4.6, where the rectified voltage is 0 for the half-period from π to 2π, and Equation 4.7 shows the average value in terms of the peak value:

$$V_{DC} = \frac{1}{T}\int_0^T v(t)\,dt = \frac{1}{2\pi}\int_0^{2\pi} V_m \sin(\omega t)\,d\theta \tag{4.6}$$

where $V_m \sin(\omega t)$ is the transformer secondary-side voltage, applied to the diode and load, neglecting the voltage drop across the diode:

$$V_{DC} = \frac{1}{2\pi}\int_0^{\pi} V_m \sin(\omega t)\,d\theta = \frac{V_m}{2\pi}\left[-\cos(\omega t)\right]_0^{\pi} = \frac{V_m}{\pi} = 0.318 V_m \tag{4.7}$$

Multiplying and dividing the second half of Equation 4.7 by $\sqrt{2}$ gives a relationship of the DC output voltage (or average) in terms of RMS of a total sine wave, as indicated in Equation 4.8. The script **halfwave_rect.m** shows the numerical computation

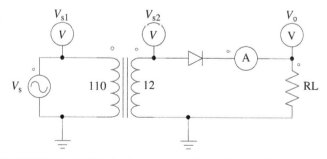

FIGURE 4.2 PSIM circuit simulation for a half-wave ideal rectifier.

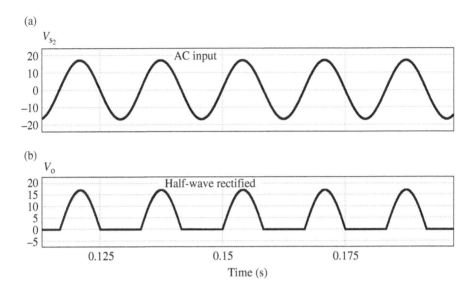

FIGURE 4.3 Output results for circuit of Figure 4.2; (a) secondary-side voltage, (b) load voltage across the resistor.

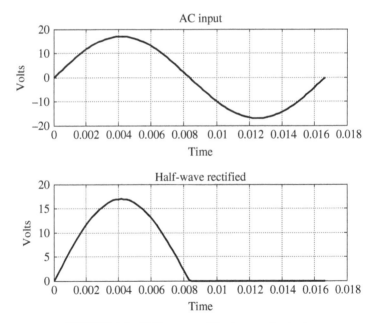

FIGURE 4.4 Half-wave rectifier input and output.

of average and effective values in MATLAB. Those values can also be compared with the PSIM simulation:

$$V_{DC} = \left(0.318\sqrt{2}\right)\frac{V_m}{\sqrt{2}} = 0.45 V_{RMS} \tag{4.8}$$

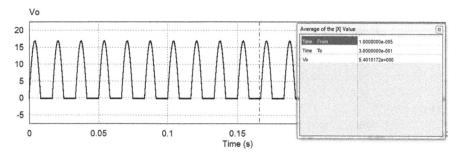

FIGURE 4.5   DC value of the load voltage across the resistor.

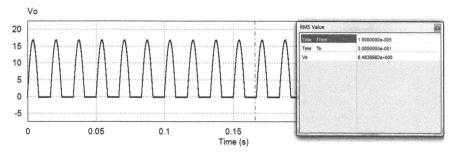

FIGURE 4.6   RMS value of the load voltage across the resistor.

```
%% halfwave_rect.m
f = 60;
T = 1/f;
Vacrms = 12;
Vm = Vacrms*1.414;
dt = T/100;
t = 0: dt: T;
vt = Vm*sin(2*pi*f*t);
vt_half = zeros(size(vt));
for n = 1: length(t)
        if vt(n) >= 0
                vt_half(n) = vt(n);
        else
                vt_half(n) = 0.0;
        end
end
row = 2;
col = 1;
figure(1), subplot(row, col, 1), plot(t, vt), grid on,
title('AC input')
xlabel('time'), ylabel('volts')
subplot(row, col, 2), plot(t, vt_half), grid on,
title('Half-wave Rectified')
```

```
xlabel('time'), ylabel('volts')
% MATLAB Numerical Integration
% Trapezoidal Integration: split the area under the curve
  into rectangles.
x = linspace(0, pi,10);    % 1.9797, gives 1 percent error
y = sin(x);
trapz(x,y)
x = linspace(0, pi,100); %1.9998 gives 0.1 percent error
y = sin(x);
trapz(x,y)
x = linspace(0, pi,1000); %2.0000
y = sin(x);
trapz(x,y)
%vt_hal
% Exact Integration to obtain
% Vdc = 0.45*Vrms = 5.4 Volts
% Vdc = 0.318*Vm  = 5.4 Volts
% Numerical Integration
%   Vdc = 5.394 volts
w = 2*pi*f;
theta = w*t;
Vdc = trapz(theta(1:50), vt_half(1:50))/(2*pi)
%%end of Matlab script -------------------------------
Vdc = 5.3940 volts
```

Figure 4.5 shows the DC value (average) of half-wave rectifier, and Figure 4.6 shows PSIM calculated RMS value.

The effective or RMS value of the output load voltage (neglecting the diode voltage drop) can be calculated by Equation 4.9:

$$V_{RMS} = \sqrt{\frac{1}{T}\int_0^T v^2(t)dt} = \sqrt{\frac{1}{2\pi}\left[\int_0^\pi \left(V_m \sin(\omega t)\right)^2 d(\omega t) + \int_\pi^{2\pi} 0 d(\omega t)\right]}$$
$$= \sqrt{\frac{V_m^2}{2\pi}\int_0^\pi \sin^2(\omega t)d(\omega t)} \tag{4.9}$$

given the trigonometric identity $\sin^2(\omega t) = \frac{1}{2}\left[1 - \cos(2\omega t)\right]$, then

$$V_{RMS} = \sqrt{\frac{V_m^2}{2\pi}\int_0^\pi \frac{1}{2}\left[1 - \cos(2\omega t)\right]d(\omega t)} = \sqrt{\frac{V_m^2}{4\pi}\left[\theta - \frac{1}{2}\sin(2\theta)\right]_{\theta=0}^{\theta=\pi}}$$

$$V_{RMS} = \sqrt{\frac{V_m^2}{4\pi}\left(\pi - 0 - 0 + 0\right)} = \frac{V_m}{2} \tag{4.10}$$

The half-wave rectifier can be characterized by several calculations dependent on the average or RMS values, such as:

- Transformer voltage and turns ratio.
- Diode rating (RMS and average current) with peak-inverse voltage.
- Forward voltage-drop can be assumed to be on the order of 0.6–0.7 V, so it is possible to compute the average diode power loss.
- Output voltage calculation (average and effective).

The average voltage is the DC value after filtering with a capacitor or a LC filter. In order to consider the voltage drop, a good approximation is

$$V_{DC} = 0.318V_m - V_{D,ON} \qquad (4.11)$$

Ripple Factor = RMS value of the AC load component/DC value of the load component and the ripple voltage is

$$\text{Ripple voltage} = \sqrt{V_{RMS}^2 - V_{DC}^2}$$

The ripple factor (in percentage) can be calculated as

$$\text{Ripple factor} = \frac{V_{RMS}}{V_{DC}} \times 100\%$$

The form factor can be calculated as

$$\text{Form factor} = \frac{V_{RMS}}{V_{DC}}$$

The peak factor is:

$$\text{Peak factor} = \frac{V_m}{V_{RMS}}$$

The authors suggest the reader to use this half-wave rectifier example to completely study the relationships that can be computed using RMS and average values for voltages and currents and observing those in PSIM as well as exporting data to MATLAB and computing the numerical evaluation with your own script. A very important ratio, as a figure of merit, is the average power divided by the apparent power at the sinusoidal source, such calculation is defined as the *system PF* (power factor).

This simple rectifier can also be implemented with a capacitive filter across the load, or an inductive–capacitive (LC) filter between the transformer and the load, and a comprehensive comparison of the average power and apparent power at the input primary side of the transformer can also be conducted.

4.3 LABORATORY PROJECT: ELECTRICAL CIRCUIT SIMULATION USING PSIM AND SIMSCAPE POWER SYSTEMS MATLAB ANALYSIS

Figure 4.8 shows the same electrical circuit used in Chapter 2 as well in Chapter 3. The circuit has been studied already in other formulation and other computer-based environments, and it is expected that the reader should be familiar already with the performance and typical responses. However, in this chapter, the emphasis is on circuit simulation, conducted initially on PSIM in this section and then on Simscape Power Systems™ (formerly SimPowerSystems™) (MATLAB toolbox). A circuit simulator is very versatile in evaluating transient responses—this electrical circuit has a current source that initially will be a pure sinusoidal, $i_x(t) = 1.5\sin(2\pi60t - 0.15)$. From there, a change in polarity will be studied by assuming that current source will flip from positive to negative, $i_x(t) = -1.5\sin(2\pi60t - 0.15)$, and then back to positive. The reader will be able to compare this circuit simulation with the block diagram-based studies of Chapter 3. Finally, the current source will be made to contain harmonics, such as $i_x(t) = [1.5\sin(2\pi60t) + 0.75\sin(2\pi180t) + 0.25\sin(2\pi300t)]$, and emphasis can be done on calculating the PF with harmonic analysis at the point of common coupling defined at where the voltage source, $v_o(t) = 169.7\cos(2\pi60t + 0.25)$, is connected.

Write a Simulation Plan—you will have to think yourself what you want to observe in the circuit, write the variables you want to capture, the time span of simulation, the initial conditions you want to implement, steady-state operation, transient conditions, variation of sources in time, and parameter variation in time. You should write a simulation plan as a document of your expectations, what you want to compare with your analysis, your theory, and what you can observe in the simulation that will help you to understand better the operation and working of the circuit or system. For example, the following items can be a possible simulation plan for the circuit on Figures 4.7 and 4.8:

- Verify the block diagram model developed previously and how it relates to the circuit.
- Investigate the effect of varying the input voltage.
- Impose small step variations.
- Study the faulty situations.

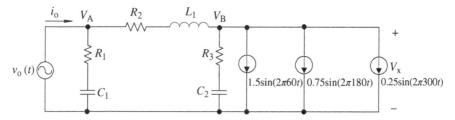

FIGURE 4.7 Electrical circuit for PSIM and SimPowerSystems simulation (current source with third and fifth harmonics).

- Perform power calculations (PF, P, and Q) and compare it to hand calculations.
- Assume that the resistance is time varying (e.g., due Joule effect) with time and capture the variations.

Figures 4.9 to 4.19 all show results and diagrams related to this laboratory project.

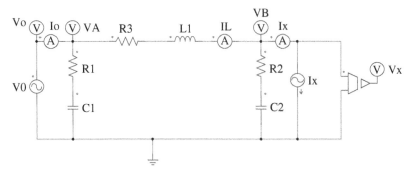

FIGURE 4.8   Electrical circuit in PSIM.

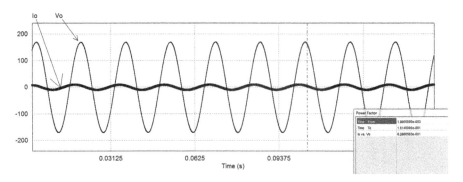

FIGURE 4.9   Power factor calculation for voltage source by PSIM.

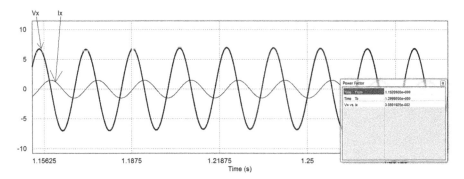

FIGURE 4.10   Power factor calculation for current source by PSIM.

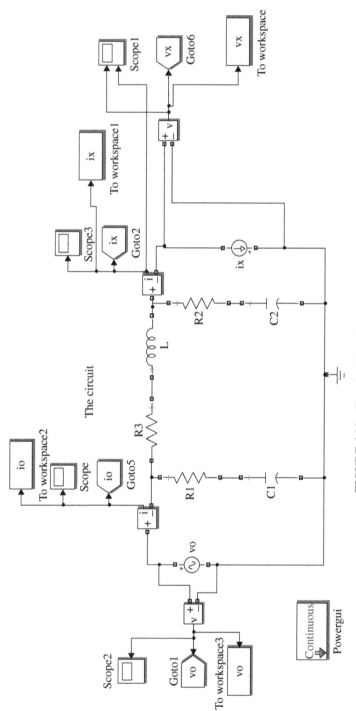

FIGURE 4.11 Electrical circuit in SimPowerSystem.

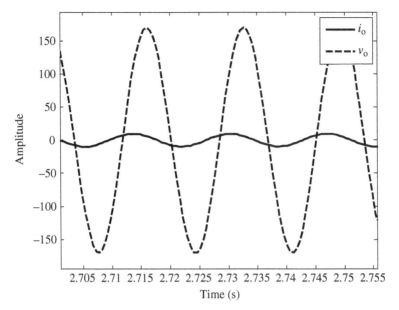

FIGURE 4.12 Current and voltage of voltage source.

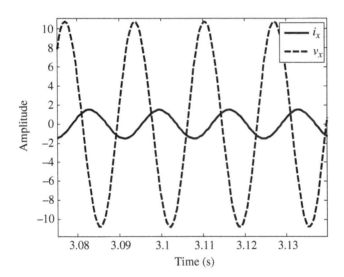

FIGURE 4.13 Current and voltage of current source.

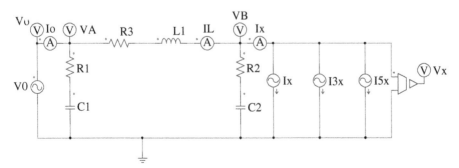

FIGURE 4.14 Electrical circuit with current source and harmonic distortion (with added third and fifth harmonics) in PSIM.

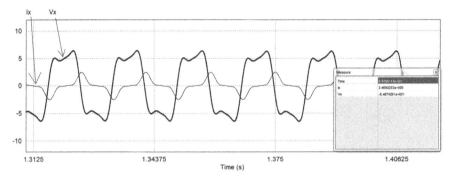

FIGURE 4.15   Power factor calculation for current source by PSIM.

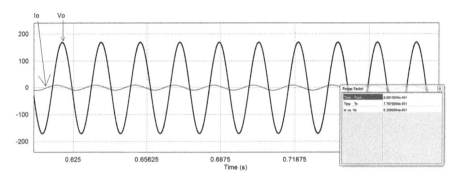

FIGURE 4.16   Power factor calculation for voltage source by PSIM.

- Compare the simulation to previous state-space analysis.
- Study the transients for $i_x(t)$ positive to negative and back to positive.
- Study the harmonic distortion for $i_x(t)$ with added third and fifth harmonics.
- Study the PF seen by voltage source $v_0(t)$ when $i_x(t)$ has added third and fifth harmonics.
- Work on this project initially in PSIM and then do the same study with Simscape Power Systems of MATLAB/Simulink in order to learn both circuit simulation environments.

Find how PSIM and SimPowerSystems can help you to find the PF. Measure the PF seen by the voltage source for several different steady-state conditions. Solve one of these conditions by hand (analytically), and show that your expected hand calculations are observed by your simulation. Export instantaneous voltage across the voltage source as well as the current, $v_0(t)$ and $i_x(t)$ for two full cycles, several points (small step-size) for several steady-state conditions. In MATLAB, we have to calculate three integrals as indicated by Equation 4.12. There are several ways to perform this average and effective computation in MATLAB, but the reader should, at this time, write a .M script that uses the function trapz in MATLAB and

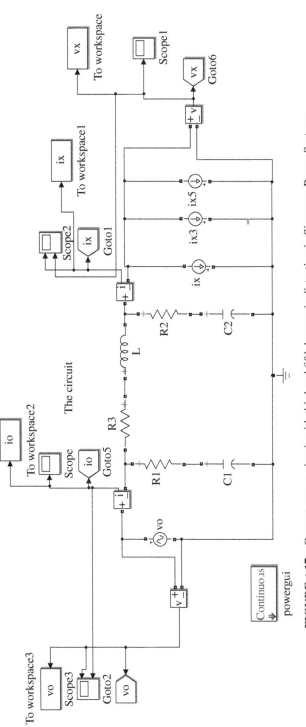

FIGURE 4.17 Current source circuit with third and fifth harmonic distortion in Simscape Power Systems.

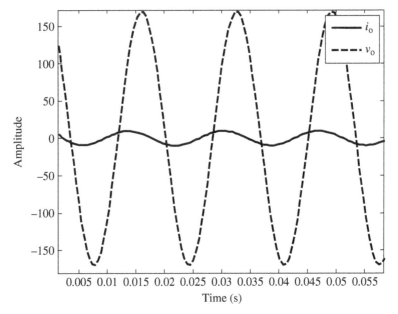

FIGURE 4.18   Current and voltage of voltage source.

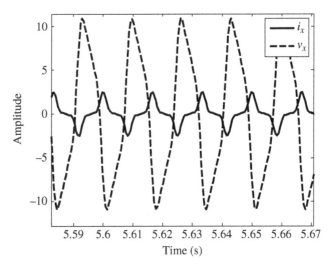

FIGURE 4.19   Current and voltage of current source.

understand who to export data from PSIM to MATLAB workspace as well as from the SimPowerSystems to MATLAB:

$$\text{PF} = \frac{\frac{1}{T}\int_0^T v(t)\cdot i(t)\,dt}{\sqrt{\frac{1}{T}\int_0^T v^2(t)\,dt}\cdot\sqrt{\frac{1}{T}\int_0^T i^2(t)\,dt}} = \frac{P_{\text{average}}}{P_{\text{apparent}}} = \frac{P_{\text{average}}}{V_{\text{RMS}}I_{\text{RMS}}} \qquad (4.12)$$

4.4 SUGGESTED PROBLEMS

1. Simulate in PSIM the reduction of the RMS harmonic distortion due to the voltage crossover in the PN junctions when using an ordinary class B push–pull amplifier as the one represented in Figure 4.20. Compare the harmonic contents of this same amplifier with a class B push–pull amplifier with an operational amplifier feedback for low load voltage applications. Do the same job for higher-voltage applications and suggest a way to reduce the harmonic contents even more.

2. The use of saturated transformers is a very common practice in electrical engineering. An overloaded transformer can produce very heavy harmonic content when making part of a power electronic source feeding a rectifier. Using the PSIM saturated star connection transformer feeding a rectifier as shown in Figure 4.21, connect a load across its output such that it causes a light (30%),

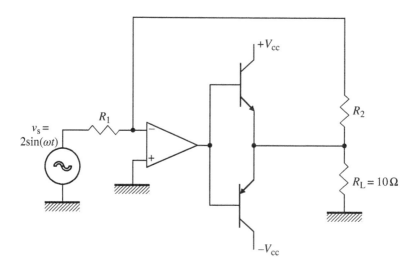

FIGURE 4.20   Class B push–pull amplifier.

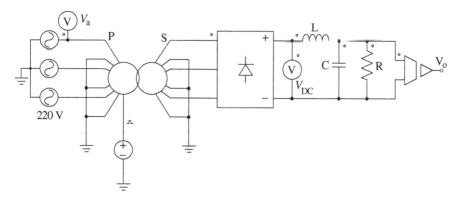

FIGURE 4.21   A star–star saturable transformer connected to a diode rectifier.

medium (90%), and heavy (150%) core saturation. What would be the percentage
harmonic increment in these three cases? What about if it was used as a delta connection? Compare these harmonic contents with the ones produced by purely
resistive and resistive–inductive loads connected directly across the transformer
terminals at the same power loadings.

3. Short circuits in distribution and transmission lines of power systems can happen
   due to several causes, such as faulted equipment, atmospheric surges, isolation
   damages by overvoltages, and many others. Such power systems faults can be
   detected and isolated by electronic circuits. Simulate in a PSIM scheme the saturated transformer as shown in Figure 4.22 below the following fault types: (i)
   short circuit between two phases, (ii) three-phase short circuit, (iii) phase-to-
   ground short circuit, (iv) open single phase, and (v) no load. What are the fault
   currents in the transformer primary in each case in order to set up the protection
   levels? What are the due overvoltages? What about the ground connection currents? Design an electronic circuit to indicate the type of fault and its severity
   levels everything ready for a protection actuation.

4. A three-phase rectifier bridge feeds a 300 V and 60 A DC load from a three-phase
   source of 415 V through a Δ-Y transformer. After a PSIM simulation, determine
   the current through each diode and the necessary specifications of the bridge
   components and transformer for a voltage drop of 0.7 V across each diode and
   uninterrupted current.

5. The semicontrolled single-phase rectifier bridge using a commutating diode as
   shown in Figure 4.23 supplies an ordinary RL load with 127 V and 60 Hz to. If the
   load is highly inductive requiring 10 A, simulate the waveforms of the voltage
   across and the current through it for $\alpha = 90°$. Suppose $L = 63$ mH and neglect all
   the other losses.

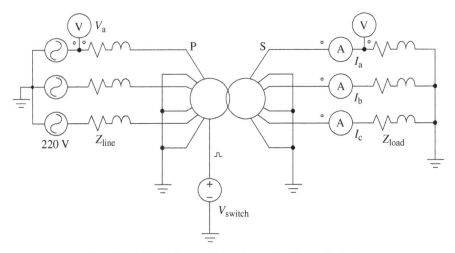

FIGURE 4.22   Saturated transformer feeding a RL load.

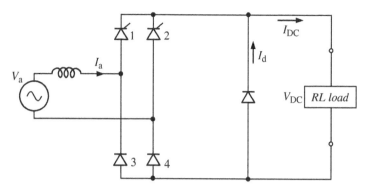

FIGURE 4.23   Semicontrolled rectifier.

FURTHER READING

BOSE, B.K, Power Electronics and AC Drives, Prentice-Hall International, Inc., Englewood Cliffs, ISBN: 0136868827, 9780136868828, 2007.

CHAKRABORTY, S., SIMÕES, M.G. and KRAMER, W.E., Power Electronics for Renewable and Distributed Energy Systems: A Sourcebook of Topologies, Control and Integration, Springer-Verlag, London, ISSN: 1865-3529, ISBN: 978-1-4471-5103-6, 2013.

CLOSE, C., FREDERICK, D.K. and NEWELL, J.C., Modeling and Analysis of Dynamic Systems, John Wiley & Sons, New York, ISBN-13: 978-0471394426, 2002.

GOLE, A.M. and FILIZADEH, S., Modelling and Simulation of Power Systems with Embedded Power Electronic Equipment, Power Electronics and Power Systems, Springer Publishers, New York/London, ISBN-10: 0387289135, ISBN-13: 978-0387289137, 2008.

HILL, W. and HOROWITZ, P., The Art of Electronics, 3rd Revised edition, Cambridge University Press, London, ISBN: 978-0521809269, 2015.

IEEE Standard on Transitions, Pulses and Related Waveforms, IEEE Std 181, 2003.

MALVINO, A.P., ZBAR, P.B. and MILLER, M.A., Basic Electronics, McGraw-Hill Companies, New York, ISBN: 007072881X, 9780070728813, 1990.

RASHID, M.H., Power Electronics: Circuits, Devices and Applications, Prentice-Hall International, Inc., Englewood Cliffs, ISBN-13: 978-0133125900, ISBN-10: 0133125904, 2007.

SLOTINE, J.J.E. and LI, W., Applied Nonlinear Control, Prentice-Hall International, Inc., Englewood Cliffs, ISBN-13: 978-0130408907, ISBN-10: 0130408905, 1991.

5

DESIGNING POWER ELECTRONIC CONTROL SYSTEMS

5.1 INTRODUCTION

Engineers must understand how to use the "analytical method" for designing and building real physical systems, instead of using "empirical methodologies," that is, building, observing, and using trial and error (which takes a lot of time and resources). Engineers should follow some steps that may lead to the best solution, consisting of:

1. Modeling
2. Setting up mathematical equations
3. Performing analysis and design based on a model
4. Setting the experimental evaluation or prototyping based on such analysis and design
5. Implementing the project
6. Final adjustments and report

Empirical methods may be expensive and even dangerous—imagine building a nuclear power plant or sending people to another planet just based on an empirical methodology. Analytical methods can be simulated in computers, and a modeling-based design may help engineers to decide if the design is satisfactory under certain

Modeling Power Electronics and Interfacing Energy Conversion Systems, First Edition.
M. Godoy Simões and Felix A. Farret.
© 2017 John Wiley & Sons, Inc. Published 2017 by John Wiley & Sons, Inc.

technical and economic considerations, in order to implement the chosen solution using physical devices.

A control system is an interconnection of components that observe the output, or internal states, and make sure that the overall system will follow a desired set point. For example, we may design control systems for (i) automatic control, such as, control of room temperature; (ii) remote control, such as, satellite orbit correction; and (iii) power amplification, that is, a control system to impress sufficient electrical power to turn on heavy electrical machines.

Control system components can be made of mechanical, electrical, hydraulic, pneumatic, chemical, or thermal systems. It can also be implemented within a computer program with input/output capabilities and software capable of analyzing the sensors, making decision, and sending signal to actuators.

When a system is in open-loop, the output has a response that is not dependent on any corrective action. For example, a water tank will fill with incoming water and it will overflow if there is no way for a maximum level that will shut off the water valve. In a closed-loop system control, the action is dependent on the output. For example, an air-conditioning system will turn off or modulate the refrigeration system based on measuring the room air temperature.

An open-loop control system can be used for very simple applications where inputs are known ahead of time and there is no disturbance, for example, the water tank could turn off the valve after timing the incoming water supply, though the output might be sensitive and can have disturbances. In such a case, an open loop would not effectively control the system. Disturbed inputs are undesirable inputs that tend to deflect the plant outputs from their desired values. They must be calibrated and adjusted at regular intervals to ensure proper operation.

Closed-loop systems depend on feedback, that is, the output is compared with the input of the system. Such error will provide appropriate control action to be formed as a function of inputs and outputs. Feedback systems add the following features to the system:

1. Reduced effect of nonlinearities and distortions
2. Increased accuracy
3. Increased system bandwidth
4. Less sensitivity to variation of system parameters
5. Reduced effects of external disturbances

However, a feedback loop may add the system a tendency toward oscillation, which is studied in control theory as "stability problem." The main objective of this chapter is to introduce how a computer simulation can be used in designing a control loop for power electronic-based systems and testing it in advance to calibrate its output. It is not possible to encompass completely such vast theory in this chapter. Therefore, we encourage the reader to study specific literature about classic control, modern control, computer-based control, digital control with proper understanding of theory of stability, mathematical analysis, and nonlinear control systems [1–3].

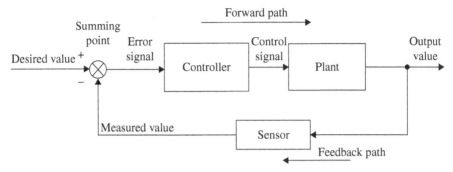

FIGURE 5.1 Closed-loop feedback control system.

A general block diagram for a closed-loop feedback control system is shown in Figure 5.1. Some important definitions related to such a control system are:

- **Reference input**—It is the actual signal input to the control system.
- **Output (controlled variable)**—It is the actual response obtained from a control system.
- **Error signal**—It is the difference between the reference input and feedback signal.
- **Closed-loop gain**—It is the amount of output signal feedback into the input to obtain the desired controlled signal.
- **Controller**—The component required to generate control signal to drive the actuator.
- **Control signal**—The signal obtained at the output of a controller is called control signal.
- **Actuator**—A power driver that produces input to the plant according to the control signal, so that output signal approaches the reference input signal.
- **Plant**—A combined system to be controlled plus the actuator.
- **Feedback**—Elements providing a mean for feeding back the output quantity in order to compare with the reference input; it can be composed of specific sensors or estimators.
- **Servomechanism**—It is usually a rotary or a linear electrical motor coupled to a mechanical system in which the controlled output is position, velocity, or acceleration.

5.1.1 Control System Design

In order to design and implement a control system, it is necessary the understanding of the following issues and topics:

- Performance specification with knowledge of the desired output values.
- Understanding of how the output changes with input, feedback sensor, resolution, and dynamic response.

- Knowledge of the controlling device.
- Knowledge of the actuating (driver) device.
- Knowledge of the plant.

Mathematical models of physical systems help to design and analyze control systems. Such systems are usually described by ordinary differential equations and can be computationally implemented by block diagrams or electrical circuits. A control system design methodology is portrayed in Figure 5.2. Unfortunately, no physical system is perfectly linear. Certain assumptions must always be made to obtain a linear model. In the presence of strong nonlinearity or in the presence of distributive effects, or time-domain parameter variation, it is not possible to obtain linear models. A common approach is to build a simplified linear model by ignoring certain nonlinearities and some physical properties that may be present in the system. Therefore, the designer must get an approximate dynamic response for the system. Then, a more complete model can be built for further analysis.

5.1.2 Proportional–Integral Closed-Loop Control

A feedback control system must be evaluated by criteria of stability determined by the closed-loop characteristic, that is, the equation

$$1 + G(s)H(s) = 0;$$

where all the roots of this equation should be in the left-hand side of the s-plane, so the inverse Laplace transformation will have only exponentials with negative coefficients or complex conjugates.

The system stability can be analyzed by graphical methods such as the *Nyquist Method* or a *Bode Method*. Therefore, if a complete mathematical formulation for the plant and control is known, the *Classical Control Theory* can be used for designing a controller that maintains the system stable with a good transient response [1, 2]. Using the frequency-domain analysis, two important design criteria are the gain margin and phase margin. The gain margin is the additional gain that a system can tolerate without a phase change that would make it unstable, that is, −180°.

The phase margin is the additional phase shift that a system can tolerate with no gain change to make it unstable, and it is defined at the frequency where the gain is unity, that is, at 0 dB. Experience shows that for satisfactory transient response, a system should have typically a phase margin greater than 35° and gain margin greater than 6 dB. When the system has a typical second-order response, the following approximate relations can be useful for evaluating the closed-loop control performance, where the bandwidth is in rad/s, the phase margin is in degrees, and the time in seconds:

$$(\text{Rise time}) \times (\text{Closed} - \text{loop bandwidth}) \cong 2.83$$

$$\% \, \text{Overshoot} \cong 75° - \text{phase margin}$$

$$\text{Damping ratio} \cong 0.01 \times \text{phase margin}$$

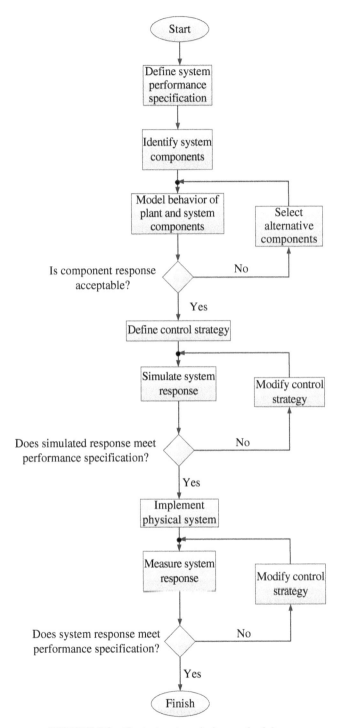

FIGURE 5.2 Control system design methodology.

Classic control supports the selection of the best controller transfer function, which could use phase-lead or phase-lag compensators, or maybe proportional–integral–derivative (PID) or proportional–integral (PI). There are typical controllers called Type I, Type II, and Type III [1, 2, 4, 5]. Those controllers are typically implemented with analog operational amplifiers with extra circuits such as for summing, subtraction, buffering, and error amplifier. In this chapter, it is presented in Section 5.2 the design of a control system for a DC/DC boost converter and a procedure for designing analog-based compensators, and then in Section 5.8, a procedure for designing a discrete PI controller is developed for a closed-loop speed control for a DC motor.

With the advent of computers and digital simulation, some intuitive understanding with trial-and-error simulation studies may help a designer to fine-tune a control loop. It is still based on the observation of response time, bandwidth, and steady-state error. A typical controller that can be easily implemented in a digital controller is the PI compensator. A PI compensator takes the error $E(s)$ of an error amplifier as input and expresses the output $Y(s)$ as the following transfer function:

$$\frac{Y(s)}{E(s)} = K_\mathrm{P} + \frac{K_\mathrm{I}}{s} = \frac{sK_\mathrm{P} + K_\mathrm{I}}{s} \tag{5.1}$$

The transfer function (5.1) can be discretized, that is, we can define a possible digital control with discretization of Equation 5.1, by assuming that this controller runs in a given sampling time T_s. We can cross multiply and expand the numerator and denominator along the two sides of the equation in order to obtain (5.2):

$$sY(s) = E(s)(sK_\mathrm{P} + K_\mathrm{I}) = sK_\mathrm{P}E(s) + K_\mathrm{I}E(s) \tag{5.2}$$

When the Laplace operator s multiplies a Laplace variable, such as $sY(s)$ or $sE(s)$, their time-domain functions are derivatives. By making the Euler difference as their first-order approximation, we arrive to Equation 5.3, where T_s is the sampling time of the digital control loop, indicated in Equation 5.4. The recurrent difference equation can be easily implemented in any DSP, microcontroller, or FPGA as in Equation 5.5:

$$\frac{y(k) - y(k-1)}{T_\mathrm{s}} = \frac{K_\mathrm{P}e(k) - K_\mathrm{P}e(k-1)}{T_\mathrm{s}} + K_\mathrm{I}e(k) \tag{5.3}$$

$$y(k) - y(k-1) = K_\mathrm{P}e(k) - K_\mathrm{P}e(k-1) + K_\mathrm{I}T_\mathrm{s}e(k) \tag{5.4}$$

$$y(k) = (K_\mathrm{P} + K_\mathrm{I}T_\mathrm{s})e(k) - K_\mathrm{P}e(k-1) + y(k-1) \tag{5.5}$$

The difference Equation 5.5 represents a discrete PI control equation. The algorithm must save the last output of the control $y(k-1)$ as well as the last error between the set point and the system variable to be controlled $e(k-1)$. It is possible to use other techniques for discretization (such as bilinear transformation), or designing an anti-windup PI controller, the reader should take a look at [6].

5.2   LABORATORY PROJECT: DESIGN OF A DC/DC BOOST CONVERTER CONTROL

The boost converter is a switching converter that has the same components as the buck converter, but it outputs a voltage greater than the input. It is a very important circuit for renewable energy-based applications, such as for photovoltaic arrays, fuel cells, and batteries. The ideal boost converter has five basic components: power semiconductor switch, a diode, an inductor, a capacitor, and a pulse width modulation (PWM) controller. The basic circuit of the boost converter is shown in Figure 5.3.

The control mechanism of the circuit in Figure 5.3 turns the switch ON and OFF, and control for how long the switch is ON in a constant frequency system, that is, using PWM. When the switch is ON, the diode does not conduct, the capacitor sustains the output voltage, the current through the inductor increases, and the energy is stored in the inductor. When the switch is OFF, the current through the inductor continues to flow via the diode and the RC network and back to the source while the inductor is discharging its energy. This converter is known as the boost converter because it can be seen that the capacitor voltage has to be higher than the source voltage; in practice, the output is limited to about four or five times the DC source. The inductor is receiving energy when the switch is closed and transferring it to the RC network when the switch is opened.

5.2.1   Ideal Boost Converter

Analysis of the circuit is carried out based on the following assumptions:

- The switch is ideal, that is, when the switch is ON, the drop voltage across it is zero and when it is open the current through it is zero.
- The diode has zero voltages drop in the conducting state and zero current in the reverse-bias mode.
- The time delays off the switch and the diode are assumed to be negligible.
- The inductor is lossless and the capacitor has a series resistance.

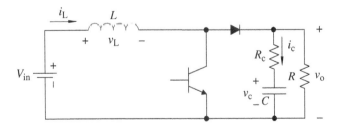

FIGURE 5.3   A regular boost converter considering the equivalent series resistance (ESR) of the capacitor.

The input and output voltage relationship of a boost converter is controlled by the duty cycle of the switch (D) according to Equation 5.6:

$$\frac{V_o}{V_i} = \frac{1}{1-D} \tag{5.6}$$

5.2.2   Small Signal Model and Deriving the Transfer Function of Boost Converter

Simplifying assumptions:

- Semiconductor devices (transistors and diode) are ideal and lossless.
- Continuous conduction mode (CCM).
- Capacitor has series and/or parallel resistance (R_c).

Parameters for a boost converter:

- Output voltage (v_o)
- Inductor current (i_L)
- Capacitor voltage (v_c)
- Input voltage (V_{in})
- Duty cycle (D)

The state variables are the ones related to the energy in the passive components, that is, the inductor current i_L and the capacitor voltage v_c. There are two different states for operating the boost converter. They will depend on whether the inductor current is always greater than zero (defining the boost converter to be in CCM). When the switch is ON, the following equivalent circuit applies (Figure 5.4):

$$L\frac{di_L}{dt} = V_{in} \tag{5.7}$$

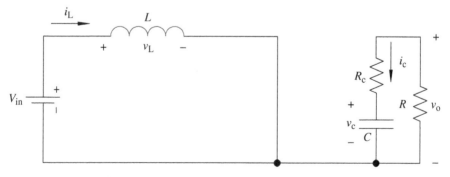

FIGURE 5.4   Boost converter when the switch is ON.

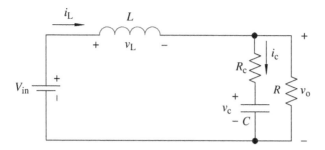

FIGURE 5.5 Boost converter when the switch is OFF.

$$C\frac{dV_c}{dt} = \frac{-V_c}{R+R_c} \to C\frac{dV_o}{dt} = \frac{-V_c R}{\left(R+R_c\right)^2} \tag{5.8}$$

and when the switch is OFF (Figure 5.5),

$$L\frac{di_L}{dt} = V_{in} - V_o \tag{5.9}$$

$$C\frac{dV_c}{dt} = \frac{R}{R+R_c}i_L - \frac{V_c}{R+R_c} \tag{5.10}$$

It is possible to average the state equations over a switching cycle by multiplying the equations for the ON system by (d) and the equations for the OFF systems by $(1-d)$. Therefore, instead of having two sets of differential equations, one when the switch is ON and another one when the switch is OFF, we can obtain an averaged model valid for the whole PWM period. In order to find these equations, we can combine two circuits using superposition: one has the inductor current with the capacitor voltage zeroed, and the other has the inductor current as open source with the capacitor voltage response. The sum of both effects leads to Equations 5.11 and 5.12:

$$L\frac{di_L}{dt} = V_{in} - \left(1-d\right)V_o \tag{5.11}$$

$$C\frac{dV_c}{dt} = \frac{\left(1-d\right)R}{R+R_c}I_L - \frac{V_c}{R+R_c} \tag{5.12}$$

In order to linearize the equations, we can assume that each circuit voltage or current contains a DC value (capitalized) with their small AC variation (lowercase with hat) such as

$$i_L = I_L + \widehat{i_L}, v_{in} = V_{in} + \widehat{v_{in}}, v_o = V_o + \widehat{v_o}, d = D + \widehat{d}, v_c = V_c + \widehat{v_c} \tag{5.13}$$

$$L\frac{d\left(I_L + \widehat{i_L}\right)}{dt} = \left(V_{in} + \widehat{v_{in}}\right) - \left(1 - D - \widehat{d}\right)\left(V_o + \widehat{v_o}\right) \tag{5.14}$$

$$C\frac{d\left(V_{c}+\widehat{v_{c}}\right)}{dt}=\frac{\left(1-D-\widehat{d}\right)R}{R+R_{c}}\left(I_{L}+\widehat{i_{L}}\right)-\frac{\left(V_{c}+\widehat{v_{c}}\right)}{R+R_{c}} \tag{5.15}$$

$$L\frac{d\widehat{i_{L}}}{dt}=\widehat{v_{in}}-\left(1-D\right)\widehat{v_{o}}+V_{o}\widehat{d} \tag{5.16}$$

$$C\frac{d\widehat{v_{c}}}{dt}=\frac{\left(1-D\right)R}{R+R_{c}}\widehat{i_{L}}-\frac{RI_{L}}{R+R_{c}}\widehat{d}-\frac{\widehat{v_{c}}}{R+R_{c}} \tag{5.17}$$

Taking the Laplace transform for these equations:

$$sL\widehat{i_{L}}\left(s\right)=\widehat{v_{in}}\left(s\right)-\left(1-D\right)\widehat{v_{o}}\left(s\right)+V_{o}\widehat{d}\left(s\right) \tag{5.18}$$

$$\left(sC+\frac{1}{R+R_{c}}\right)\widehat{v_{c}}\left(s\right)=\frac{\left(1-D\right)R}{R+R_{c}}\widehat{i_{L}}\left(s\right)-\frac{RI_{L}}{R+R_{c}}\widehat{d}\left(s\right) \tag{5.19}$$

$$V_{o}=V_{c}+R_{c}C\frac{dV_{c}}{dt}\rightarrow\frac{V_{o}\left(s\right)}{1+sR_{c}C}=V_{c}\left(s\right) \tag{5.20}$$

From (5.19):

$$\widehat{i_{L}}\left(s\right)=\frac{R+R_{c}}{\left(1-D\right)R}\left(\left(sC+\frac{1}{R+R_{c}}\right)\widehat{v_{c}}\left(s\right)+\frac{RI_{L}}{R+R_{c}}\widehat{d}\left(s\right)\right) \tag{5.21}$$

From (5.18) and (5.19):

$$sL\frac{R+R_{c}}{\left(1-D\right)R}\left[\left(sC+\frac{1}{R+R_{c}}\right)\widehat{v_{c}}\left(s\right)+\frac{RI_{L}}{R+R_{c}}\widehat{d}\left(s\right)\right]=\widehat{v_{in}}\left(s\right)-\left(1-D\right)\widehat{v_{o}}\left(s\right)+V_{o}\widehat{d}\left(s\right)$$

$$\tag{5.22}$$

The transfer function that is useful for control is the one that relates the output voltage variation with the duty cycle variation, that is, as given by Equation 5.23:

$$G\left(s\right)=\frac{\widehat{v_{o}}\left(s\right)}{\widehat{d}\left(s\right)}\bigg|_{\widehat{v_{in}}=0} \tag{5.23}$$

From (5.20) and (5.22):

$$\left[\frac{\frac{sL}{1-D}\left(sC+\frac{1}{R+R_{c}}\right)}{1+sR_{c}C}+\left(1-D\right)\right]\widehat{v_{o}}\left(s\right)=\left[\frac{V_{in}}{1-D}-\left(\frac{RI_{L}}{R+R_{c}}\right)\frac{sL}{1-D}\right]\widehat{d}\left(s\right) \tag{5.24}$$

$$\frac{\widehat{v_o}(s)}{\widehat{d}(s)} = \frac{\left(\dfrac{V_{in}}{1-D} - \left(\dfrac{RI_L}{R+R_c}\right)\dfrac{sL}{1-D}\right)(1+sR_cC)}{\dfrac{sL}{1-D}\left(sC + \dfrac{1}{R+R_c}\right) + (1-D)(1+sR_cC)} \tag{5.25}$$

From the DC/DC boost converter analysis, we have that

$$I_L = \frac{I_o}{1-D} = \frac{V_o}{R(1-D)} = \frac{V_{in}}{R(1-D)^2} \tag{5.26}$$

The control transfer function for the boost converter in CCM is given by Equation 5.27:

$$\frac{\widehat{v_o}(s)}{\widehat{d}(s)} = \frac{V_{in}}{(1-D)^2}\left(1 - s\frac{L_e}{R}\right)\frac{(1+sR_cC)}{L_eC\left(s^2 + s\left(\dfrac{1}{RC} + \dfrac{R_c}{L_e}\right) + \dfrac{1}{L_eC}\right)} \tag{5.27}$$

where

$$L_e = \frac{L}{(1-D)^2} \tag{5.28}$$

5.2.3 Control Block Diagram and Transfer Function

The PWM circuit converts the output from the compensated error amplifier to a duty ratio. The error amplifier output voltage v_c is compared to a sawtooth waveform with amplitude v_p. Chapter 6 covers electronics and circuits used for implementation of analog pulse-width modulation (PWM) and sinusoidal PWM (SPWM). The output of the PWM circuit is high, while v_c is larger than the sawtooth and is zero when v_c is less than the sawtooth. If the output voltage falls below the reference, the error between the converter output and the reference signal increases, causing v_c to increase and the duty ratio to increase. Conversely, a rise in output voltage reduces the duty ratio. A transfer function for the PWM process is given by Equation 5.29:

$$D = \frac{V_c}{V_p} \tag{5.29}$$

The output is high when v_c from the compensated error amplifier is higher than the sawtooth waveform. The transfer function of the PWM circuit is given by Equation 5.30:

$$\frac{d(s)}{v_c(s)} = \frac{1}{V_p} \tag{5.30}$$

5.2.3.1 *Type III Compensator* Type III amplifier is shown in Figure 5.6. The Type III amplifier provides an additional phase angle boost compared to the Type II circuit and is used when an adequate phase margin is not achievable using the Type II amplifier. The small-signal transfer function is expressed in terms of input and feedback impedances Z_i and Z_f:

$$G(s) = \frac{\tilde{V}_c(s)}{\tilde{V}_o(s)} = -\frac{Z_f}{Z_i} = -\frac{\left(R_2 + \dfrac{1}{sC_1}\right) \middle\| \dfrac{1}{sC_2}}{R_1 \middle\| \left(R_3 + \dfrac{1}{sC_3}\right)} \tag{5.31}$$

The zeros and poles for this Type III compensator are

$$\omega_{z_1} = \frac{1}{R_2 C_1} \tag{5.32}$$

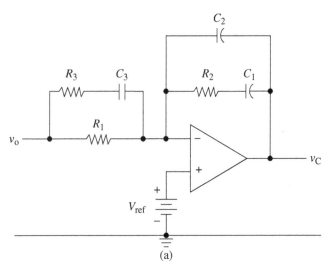

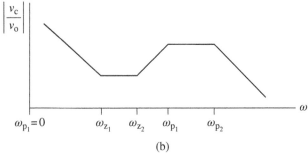

FIGURE 5.6 Controller frequency response. (a) Type III compensated error amplifier; (b) Bode magnitude plot.

$$\omega_{z_2} = \frac{1}{(R_1 + R_3)C_3} \approx \frac{1}{R_1 C_3} \tag{5.33}$$

$$\omega_{p_1} = 0 \tag{5.34}$$

$$\omega_{p_2} = \frac{C_1 + C_2}{R_2 C_1 C_2} \approx \frac{1}{R_2 C_2} \tag{5.35}$$

$$\omega_{p_3} = \frac{1}{R_3 C_3} \tag{5.36}$$

$$
\begin{aligned}
\theta_{comp} &= -180° + \tan^{-1}\left(\frac{\omega}{\omega_{z_1}}\right) + \tan^{-1}\left(\frac{\omega}{\omega_{z_2}}\right) - 90° - \tan^{-1}\left(\frac{\omega}{\omega_{p_2}}\right) - \tan^{-1}\left(\frac{\omega}{\omega_{p_3}}\right) \\
&= -270° + \tan^{-1}\left(\frac{\omega}{\omega_{z_1}}\right) + \tan^{-1}\left(\frac{\omega}{\omega_{z_2}}\right) - \tan^{-1}\left(\frac{\omega}{\omega_{p_2}}\right) - \tan^{-1}\left(\frac{\omega}{\omega_{p_3}}\right)
\end{aligned}
\tag{5.37}
$$

5.3   DESIGN OF A TYPE III COMPENSATED ERROR AMPLIFIER

There are two suggested procedures for designing the controller, (i) the K Method and (ii) pole placement, discussed next.

5.3.1 K Method

The K factor method [5] can be used for designing the Type III controller; in this method, the zeros are placed at the same frequency to form a double zero, and the second and third poles are placed at the same frequency to form a double pole:

$$\omega_z = \omega_{z_1} = \omega_{z_2} \tag{5.38}$$

$$\omega_p = \omega_{p_2} = \omega_{p_3} \tag{5.39}$$

$$R_2 = \frac{|G(j\omega_{co})| R_1}{\sqrt{K}} \tag{5.40}$$

$$C_1 = \frac{\sqrt{K}}{\omega_{co} R_2} = \frac{\sqrt{K}}{2\pi f_{co} R_2} \tag{5.41}$$

$$C_2 = \frac{1}{\omega_{co} R_2 \sqrt{K}} = \frac{1}{2\pi f_{co} R_2 \sqrt{K}} \tag{5.42}$$

$$C_3 = \frac{\sqrt{K}}{\omega_{co} R_1} = \frac{\sqrt{K}}{2\pi f_{co} R_1} \qquad (5.43)$$

$$R_3 = \frac{1}{\omega_{co} C_3 \sqrt{K}} = \frac{1}{2\pi f_{co} C_3 \sqrt{K}} \qquad (5.44)$$

$$K = \tan\left(\frac{\theta_{comp} + 90^0}{4}\right)^2 \qquad (5.45)$$

$$\theta_{comp} = \theta_{phase\,margin} - \theta_{converter} \qquad (5.46)$$

5.3.2 Poles and Zeros Placement in the Type III Amplifier

As an alternative to the K factor method described previously, some designers place the poles and zeros of the Type III amplifier at specified frequencies. In placing the poles and zeros, a frequency of particular interest is the resonant frequency of the LC filter in the converter. Neglecting any resistance in the inductor and capacitor,

$$\omega_{LC} = \frac{1}{\sqrt{LC}} \rightarrow f_{LC} = \frac{1}{2\pi\sqrt{LC}} \qquad (5.47)$$

The first zero is commonly placed from 50 to 100% of f_{LC}, the second zero is placed at f_{LC}, the second pole is placed at the ESR zero in the filter transfer function $\left(\frac{1}{r_C C}\right)$, and the third pole is placed at one half the switching frequency. Table 5.1 indicates placement of the Type III error amplifier poles and zeros.

TABLE 5.1   Placement of Type III Error Amplifier Poles and Zeros

	Expression	Placement
First zero	$\omega_{z_1} = \dfrac{1}{R_2 C_1}$	50–100% of ω_{LC}
Second zero	$\omega_{z_2} = \dfrac{1}{(R_1 + R_3)C_3} \approx \dfrac{1}{R_1 C_3}$	At ω_{LC}
First pole	$\omega_{p_1} = 0$	—
Second pole	$\omega_{p_2} = \dfrac{C_1 + C_2}{R_2 C_1 C_2} \approx \dfrac{1}{R_2 C_2}$	At the ESR zero $= \dfrac{1}{r_C C}$
Third pole	$\omega_{p_3} = \dfrac{1}{R_3 C_3}$	At one half the switching frequency, $\dfrac{\omega_{sw}}{2}$

5.4 CONTROLLER DESIGN

Type III controller can be used to compensate the output voltage of a DC/DC boost converter, which varies with changes in the input voltage as well as the output load. There are two methods to design Type III controller, and this section emphasizes the manual placement of zeros and poles.

The behavior of a boost converter transfer function must be analyzed from the open-loop stability analysis viewpoint. To do this, all of the parameters of the boost converter must be defined, and by using the converter transfer function, we can run simulations for several conditions to plot the frequency response (Bode plot), for example, using MATLB. We can assume that the gain of PWM is equal to 1 (peak-to-peak voltage of the sawtooth on the PWM analog comparator) and also assume a gain of the voltage divider, which depends on the output voltage ratio to the reference voltage, considered here as $G_s = 0.0132$. In MATLAB, we can use the "margin" command to observe the Bode plot of any transfer function with their phase margin and gain margin. In this project, we see that phase margin of our system is about $9°$, and our goal would be to increase it to about $30°$ (Figure 5.7).

The boost converter parameters are indicated in the following MATLAB script lines. The duty cycle is calculated for the operating conditions:

$$V_o = \frac{V_{in}}{1-D} \quad \underset{}{V_{in} = 80\,V} \quad \text{and } V_o = 380\,V \quad \underset{}{D = 0.789} \tag{5.48}$$

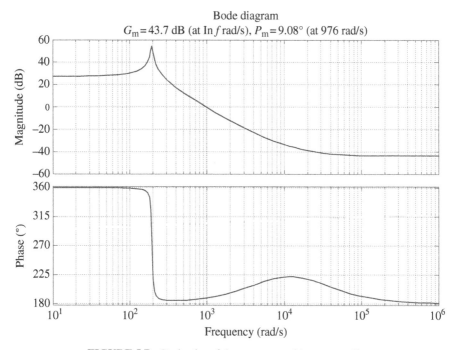

FIGURE 5.7 Bode plot of the converter without controller.

```
%%Parameters of the boost converter
V_in = 80; % Input Voltage
V_out = 380; %Output Voltage
L = 250e-6;
C = 4700e-6;
D = 300/380; % Duty cycle
R= 150; %Load Resistance
r = 41e-3; % Capacitor Resistance
Le = L/((1-D)^2); % For Boost converters
```

The value of the resonant frequency $f_{LC} = \dfrac{1}{2\pi\sqrt{LC}}$ of this boost converter will be equal to 146.8254 Hz. Therefore, the value of ω_{LC} is equal to 922.5334 rad/s. By assuming that the PWM switching frequency is 20 kHz, we can use the suggested placement for zeros and poles of the Type III compensator in Table 5.2.

If it is used the second method, we need to make assumptions about the values of two parameters, for example, common choices could be for R_1 and C_2. It is known that C_2 should be much smaller than C_1 so one can chose a value for C_2, for example, 1.6 nF, and consider R_1 to be 1 kΩ. Based on previous equations, Table 5.3 lists the values of several parameters of the compensator.

Using the equations for the compensator transfer function, we can simulate it in MATLAB. This has been done using a structure variable named "G" in the simulation file, and then using the "series" command, we can find the whole system open-loop transfer function, which has a structured variable named "GH" in the simulation file.

Eventually, the Bode diagram of the boost converter, the compensator, and the whole control system can be plotted. Figure 5.8 shows frequency analysis diagrams. As we can observe from the Figure 5.8, the controller is working quite well, and in the crossover frequency, which is about 1.2 kHz, the phase margin is 31°.

TABLE 5.2   Suggested Placement for Zeros and Poles of the Type III Compensator

$\omega_{z_1} = 461.2667$
$\omega_{z_2} = 922.5334$ rad/s
$\omega_{p_1} = 0$
$\omega_{p_2} = 5189.4$
$\omega_{p_3} = 62\,832$

TABLE 5.3   The Values of Different Parameters of the Compensator

$C_1 = 0.18$ nF	$C_3 = 1.084\,\mu\text{F}$
$R_2 = 0.12$ MΩ	$R_3 = 14.6825\,\Omega$
$R_a = 1000\,\Omega$	$R_b = 13.33\,\Omega$

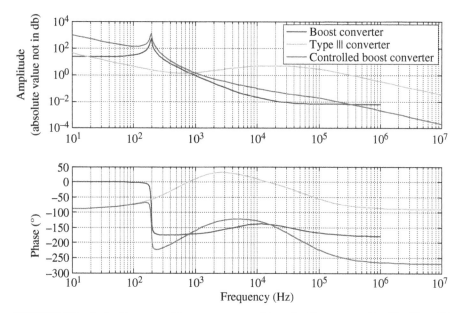

FIGURE 5.8   Bode diagrams of boost converter with and without controller. The Type III compensator bode diagram is also plotted.

5.5   PSIM SIMULATION STUDIES FOR THE DC/DC BOOST CONVERTER

A circuit diagram for the boost converter in open loop (without a controller) is shown in Figure 5.9. The output voltage without changing in input is portrayed in Figure 5.10. The initial transient occurs because all the initial conditions are zero.

By changing the input voltage from 80 to 96 V with using two-step voltage source at 0.6 s, the output voltage of boost converter changes from 400 to 480 V. The value of resistance of the capacitor is 41 mΩ, and the duty cycle is 0.789. Figure 5.11 shows the output voltage when the input increases at 0.6 s, without any controller, that is, open-loop response.

5.6   BOOST CONVERTER: AVERAGE MODEL

Figure 5.12 shows the circuit for this design, where dependent current and voltage sources are used, instead of using switches, such as a real transistor and a diode. Therefore, this simulation can run with larger step size and responds to the average values instead of the instantaneous switching variable responses. Figure 5.13 shows the output of boost converter without changes in input and load.

Figure 5.14 shows output voltage of boost converter when the input voltage changes from 80 to 96 V at 0.6 s.

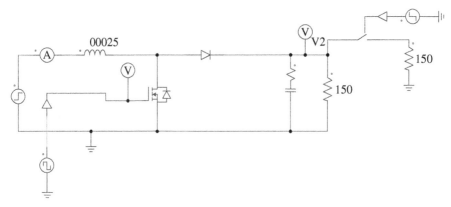

FIGURE 5.9 Boost converter in open loop.

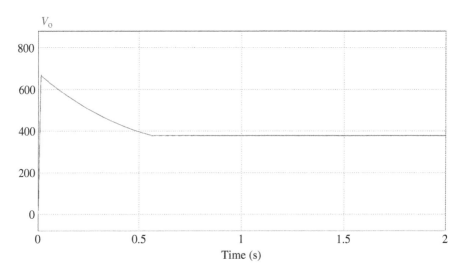

FIGURE 5.10 Output voltage without changing input and without controller.

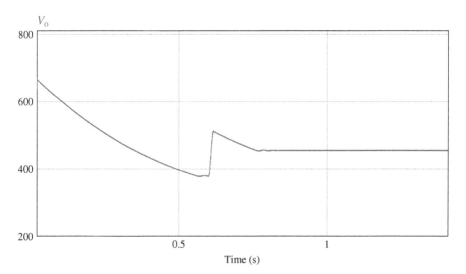

FIGURE 5.11 Output voltage when the input increases at 0.6 s and without controller.

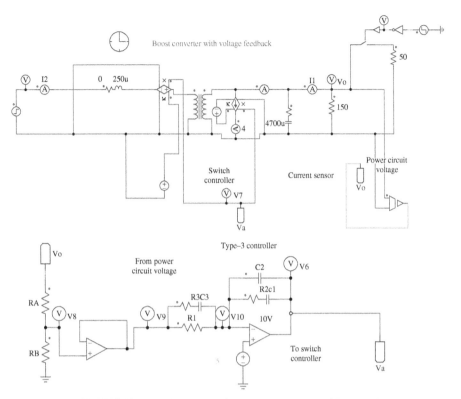

FIGURE 5.12   Circuit diagram for a boost converter with controller.

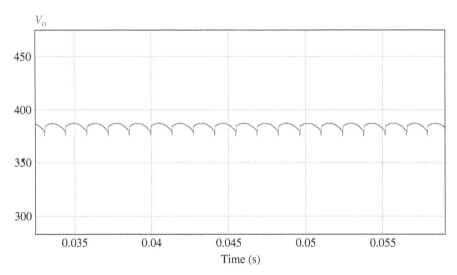

FIGURE 5.13   Output voltage without changing in input and load with controller.

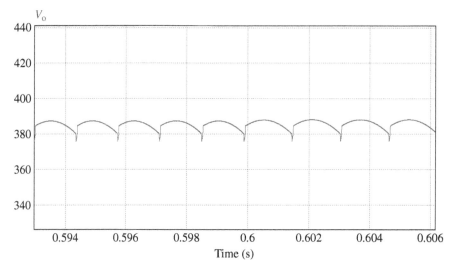

FIGURE 5.14   Output voltage when input voltage increases from 80 to 96 at 0.6 s (boost converter with controller).

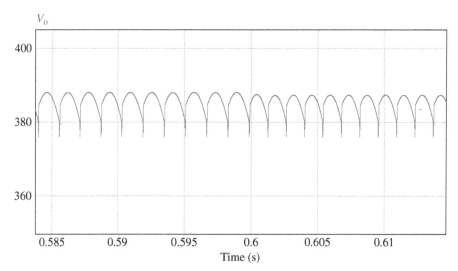

FIGURE 5.15   Output voltage when input voltage decreases from 96 to 80 V at 0.6 s (boost converter with controller).

As depicted in Figures 5.13, 5.14, and 5.15, the controller works well. At 0.6 s, there is an increasing input voltage from 80 to 96 V. There are no changes in V_o, and it remains at 380 V. The input voltage changes from 96 to 80 V at 0.6 s, and Figure 5.15 shows the controller response.

Figure 5.16 shows the output voltage of the boost converter when its load changes from 300 to 150 Ω by using bidirectional switch at 0.8 s ($V_{in} = 80$ V), and

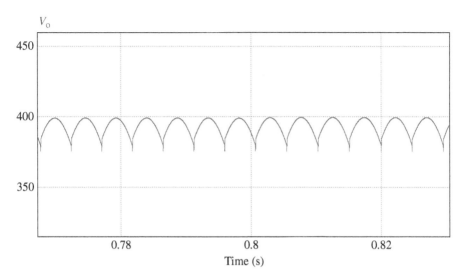

FIGURE 5.16 Output voltage with decreasing R_L from 300 to 150 Ω at 0.8 s and with controller ($V_{in} = 80$ V).

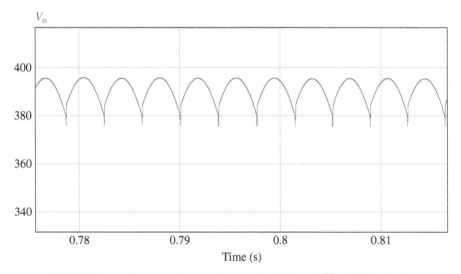

FIGURE 5.17 Output voltage with decreasing R_L from 150 to 300 Ω at 0.8 s.

Figure 5.17 shows the output voltage of boost converter when the load changes from 150 to 300 Ω.

5.7 FULL CIRCUIT FOR THE DC/DC BOOST CONVERTER

Figure 5.18 shows the PSIM simulation for a full switching circuit with a transistor, a diode, and operational amplifiers, and a sawtooth waveform for the PWM is implemented. Figures 5.19 and 5.20 show the output voltage of the boost

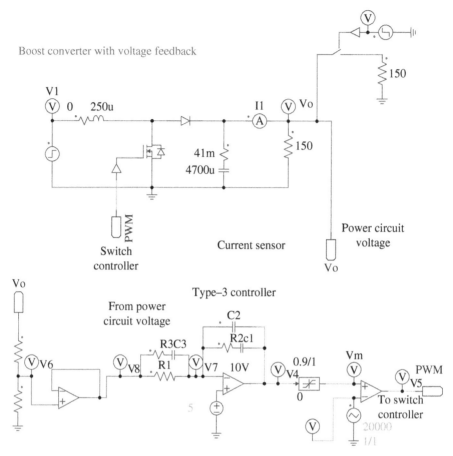

FIGURE 5.18   Boost converter with full switching circuit.

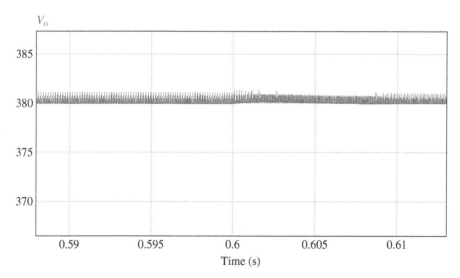

FIGURE 5.19 . Output voltage when input voltage increases from 80 to 96 V at 0.6 s (boost converter with controller).

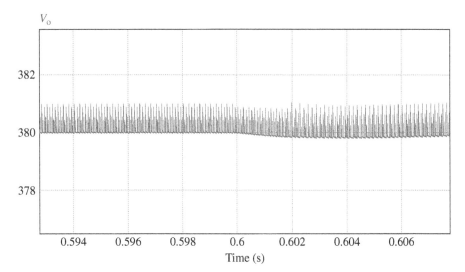

FIGURE 5.20   Output voltage when input voltage decreases from 96 to 80 V at 0.6 s (boost converter with controller).

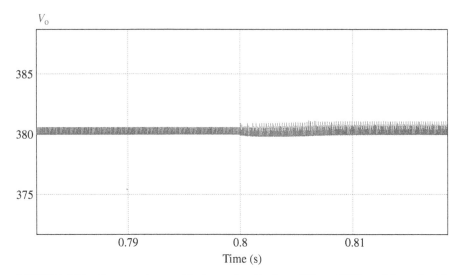

FIGURE 5.21   Output voltage with decreasing R_L from 300 to 150Ω at 0.8 s and with controller (V_{in} is 80 V).

converter when the input voltage changes initially from 80 to 96 V and from 96 to 80 V at 0.6 s, respectively.

It is possible to observe a very good controller response with respect to changes in input voltage. For example, at the instant 0.6 s with increasing input voltage from 80 to 96 V, there is no change in the output voltage (V_o) and it remains at 380 V. Figure 5.21 shows the output voltage of the boost converter when the load changes from 300 to

150 Ω by using a bidirectional switch at 0.8 s (V_{in} = 80 V). Figure 5.22 shows the output voltage of boost converter when the load changes from 150 to 300 Ω. In order to show a good performance of a controller, the input voltage at 0.6 s and the output resistor at 0.7 s change. As observed in Figure 5.23, the controller responded very well to these disturbances.

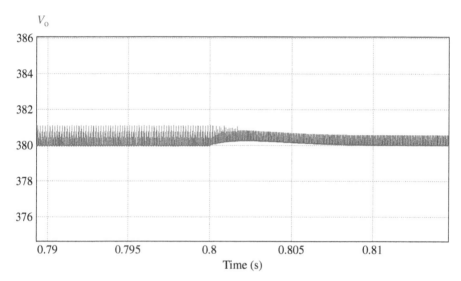

FIGURE 5.22   Output voltage with decreasing R_L from 150 to 300 Ω at 0.8 s.

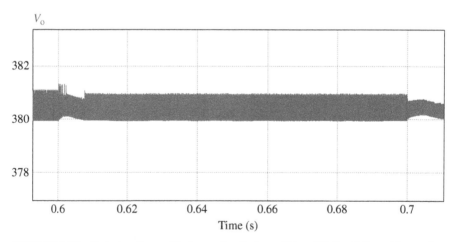

FIGURE 5.23   Output voltage with decreasing R_L from 150 to 300 Ω at 0.7 s and increasing input voltage from 80 to 96 V at 0.6 s.

5.8   LABORATORY PROJECT: DESIGN OF A DISCRETE CONTROL IN MATLAB CORUNNING WITH A DC MOTOR MODEL IN Simulink

The goal of this project is to implement a DC motor model in Simulink and a controller in MATLAB in order to assess a digital control scheme. Initially, a closed-loop control is studied in the Simulink environment, so as some preliminary response and debugging would help the student to learn how the system works. In a computational model-based design methodology, it is always important to simulate the system in a familiar approach first before conducting other programming or modeling techniques considered to be new at next stage.

After developing the DC motor model in Simulink and evaluating a PI speed control loop, the controller is discretized and implemented in MATLAB. However, we run the simulations in MATLAB, calling the Simulink model from the MATLAB workspace. This is a powerful way to integrate MATLAB-based controllers with all the mathematics available at the MATLAB workspace, that is, at the prompt ≫, and precisely control the sampling time. This approach can also be implemented in another way, the Simulink-based model might be running, and a function call is made to run a function with MATLAB. Another possibility is to run the discrete control inside C-callable functions that would then be debugged for hardware implementation. However, this chapter does not cover all these further possibilities. This content is outside of the scope of this book. We encourage enquiring readers to explore further possibilities of co-simulation of MATLAB and Simulink aiming real hardware prototyping.

We developed a strategy that allows the Simulink model to be called within MATLAB and run. For each and every step size MATLAB will call Simulink in order to obtain the angular speed of the DC motor (from the Simulink model). This is the input in the discretized equation of the PI controller. The output of such controller becomes the input for the model. This procedure is cycled with a **FOR** loop in a MATLAB script; in this **FOR** loop, we use the **SIM** function which calls a Simulink file to run in MATLAB.

With such strategy, it is possible to use the whole MATLAB environment. This approach makes a very powerful simulation methodology, as the plant can be executed with a block diagram in Simulink. In addition, it is possible to export and import data from other sources, available in the MATLAB workspace, and by changing the step size of the MATLAB script, it is possible to study the effect of a real digital control algorithm and how the sampling time T_s is related to the controller gains and the overall closed-loop response. Figure 5.24 shows the DC permanent magnet machine model made in Simulink. The modeling equations, using the Laplace domain, are

$$v_a(s) = R_a I_a(s) + sL_a I_a(s) + K_E \omega(s) \tag{5.49}$$

$$T_e(s) = K_T I_a(s) \tag{5.50}$$

$$T_e(s) - T_L(s) = Js\omega(s) \tag{5.51}$$

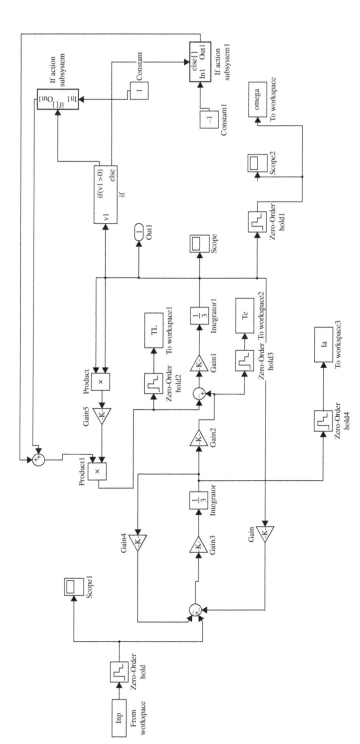

FIGURE 5.24 DC motor Simulink model.

The load torque is considered proportional to the square of the shaft angular speed, typical of fan, or bladed, or ventilation-based loads:

$$T_L(s) = K_\omega \omega^2(s) \tag{5.52}$$

The MATLAB workspace can retrieve different parameters such as I_a, ω, T_L, and T_e. The workspace can get the output variables from workspace to Simulink as an input. A MATLAB script is indicated next, where the command "**Sim**" is executed inside a **FOR** loop, and the PI controller is a difference equation inside the loop.

The goal of this project was to interact a Simulink model with a MATLAB program, where the DC machine model was made in Simulink (file **dcmotor**), whereas the PI controller is discrete and running in MATLAB (file DCmachine.M), both will work together; for each simulated step size, MATLAB calls the model in Simulink (with command **sim**) and runs the plant model. There is an error computation by subtracting the machine angular speed from reference speed, and then the PI equation is calculated:

```
sim('dcmotor01')
E(K)=wref-omega(K);
Y(K)=(KP+KI*Ts)*E(K)-KP*E(K-1)+Y(K-1);
```

The output of the controller is made as input for the model. This procedure is made continuous with a FOR loop. The control sampling time is 0.01 s; the initial values are loaded in the variable x Initial and the index K cycles from one step to next one as:

```
Ts=0.01;
xInitial=[0 0]
for K=2:1:2001
```

This code can be used to implement any digital controller and to study the effect of any variation of gains, control parameters, discretization method, effect of discretization in the overall close-loop response, and any details required for designing a digital control.

```
%%Design of a Discrete Control in Matlab Running with a
DC Motor Model in Simulink
clear all
clc
KP=0.1;
Y(1)=0;
KI=10;
omega(2)=0;
wref=80;
W=0;
```

```
a=0;%Te
b=0;%TL
d=0;%Ia
e=0;%Va
t=0;
E(1)=0
Inp=[0,0];
Ts=0.01;
xInitial=[0 0]
for K=2:1:2001
      sim('dcmotor');
      E(K)=wref-omega(K);
      Y(K)=(KP+KI*Ts)*E(K)-KP*E(K-1)+Y(K-1);
%limiter
  if Y(K)>110;
              Y(K)=110 ;
  if Y(K)<-110 ;
              Y(K)=-110;
  end
end
      Inp=[t,Y(K)];
      C=omega(K);
      W=[W,C];
      TE=Te(K);
      a=[a TE];
      Tl=TL(K);
      b=[b Tl];
      IA=Ia(K);
      d=[d IA];
      Va=Y(K);
      e=[e Va];
      t=t+Ts;
if t > 10
       wref = 20;
elseif t > 5
       wref = -80;
else
       wref = 90;
end
  status = [t wref]
end
t=0:0.01:20
s(1)=subplot(5,1,1);
plot(s(1),t,a);ylabel('Te')
grid on
```

```
s(2)=subplot(5,1,2);
plot(s(2),t,b);ylabel('TL')
grid on
s(3)=subplot(5,1,3);
plot(s(3),t,d);ylabel('Ia')
grid on
s(4)=subplot(5,1,4);
plot(s(4),t,e);ylabel('Va')
grid on
s(5)=subplot(5,1,5);
plot(s(5),t,W);ylabel('w')
grid on
```

In order to observe the transient response, the angular shaft reference speed changes three times in three different instants, as indicated by Figure 5.25, showing the armature current I_a, the load torque T_L, and the electrical machine torque T_e, respectively. The input voltage v_a, which is provided by the MATLAB script, is portrayed in Figure 5.25. We can observe in the figures that load torque and machine torque follow the equation $T_e(s) - T_l(s) = Js\omega(s)$; so when the angular speed $\omega(s)$ is constant, the derivative of $\omega(s)$ is zero and $T_e(s) = T_l(s)$.

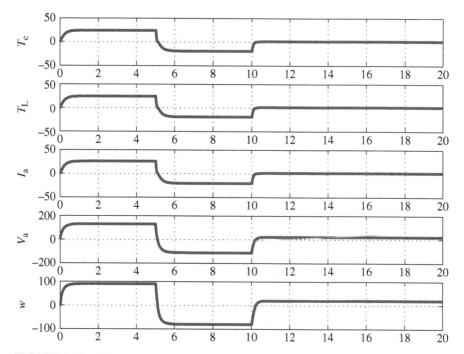

FIGURE 5.25 Electrical torque, load torque, armature current, armature voltage, and shaft angular speed.

5.9 SUGGESTED PROBLEMS

1. The circuits in Figures 5.26 (a) and (b) are typical circuits used for filtering and also closing the loop of a plant. The transfer function is the ratio of Laplace transform of the output variable to the Laplace transform of the input variable assuming zero initial conditions:

   i. Derive (by hand) the transfer functions (indicated in Figure 5.26).

   ii. Plot their Bode diagrams and find out their frequency response characteristics.

   iii. Design an op-amp-based circuit that has the same transfer function and will not suffer of loading and buffering effects that a passive circuit has.

   iv. Design a digital low-pass filter equivalent to Figure 5.26a.

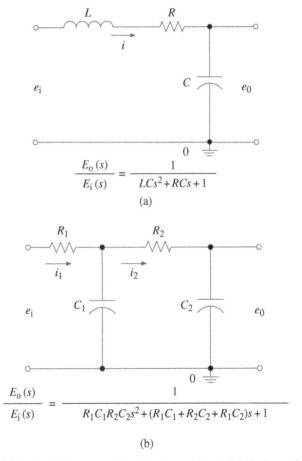

$$\frac{E_o(s)}{E_i(s)} = \frac{1}{LCs^2 + RCs + 1}$$

(a)

$$\frac{E_o(s)}{E_i(s)} = \frac{1}{R_1C_1R_2C_2s^2 + (R_1C_1 + R_2C_2 + R_1C_2)s + 1}$$

(b)

FIGURE 5.26 Typical low-pass filter circuits used in closing the loop of systems.

     v. Using Simulink, MATLAB, and PSIM, implement a noisy voltage source with several harmonics. Use the transfer function, the circuits (passive and op-amp active ones), plus the digital filter to obtain a clean filtered signal. Compare and make analysis of all those possible filter solutions.

2. A mechanical translational system and a rotational one are indicated in Figure 5.27a and b. Implement their models in Simulink. Understand the relationship of the variables, where:

Translational Movement	Rotational Movement
Linear displacement y	Angular displacement θ
Force u	Torque T
Mass m	Moment of inertia J

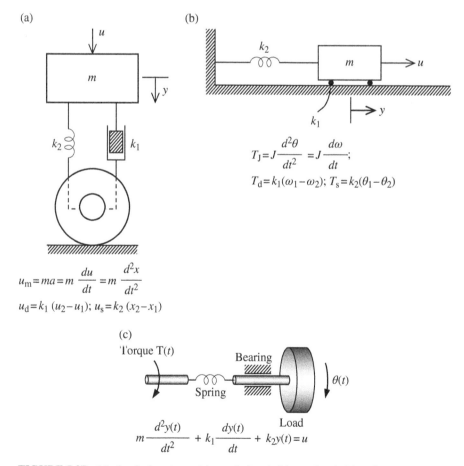

(a)

(b)

$$T_J = J\frac{d^2\theta}{dt^2} = J\frac{d\omega}{dt};$$

$$T_d = k_1(\omega_1 - \omega_2); \quad T_s = k_2(\theta_1 - \theta_2)$$

$$u_m = ma = m\frac{du}{dt} = m\frac{d^2x}{dt^2}$$

$$u_d = k_1(u_2 - u_1); \quad u_s = k_2(x_2 - x_1)$$

(c)

Torque $T(t)$

Bearing

Spring

Load

$$m\frac{d^2y(t)}{dt^2} + k_1\frac{dy(t)}{dt} + k_2y(t) = u$$

FIGURE 5.27   Mechanical systems: (a) translational, (b) rotational, (c) spring-mass system.

    i. Using Simulink, close the loop for both systems in velocity (or angular speed) and in position.

    ii. Using Simulink, design a digital control for both systems in velocity (or angular speed) and in position.

    iii. Run the plants in Simulink and their digital controllers in MATLAB.

3. Show that the mechanical system in Figure 5.28a is mathematically analogous to the electrical system in Figure 5.28b. Find good parameters for the mechanical system, translate these parameters into electrical circuit components, and simulate (a) with Simulink and (b) with PSIM. The simulation studies should be compatible for a broad range of operation and settings.

    i. Develop a transfer function model for the equivalent circuit in Figure 5.28b; develop a controller for one of the variables, for example, the capacitor voltage V_{C_3}; close the loop; and after the simulation in PSIM is fully operational, use the same controller in Simulink for controlling the mass M_3. This project of a controller in the electrical circuit domain will support the implementation of a controller in the mechanical system domain.

4. Figure 5.29 shows a three-phase voltage source inverter connected to a three-phase RL load with a DC input voltage, with SPWM analog control with a

(a)

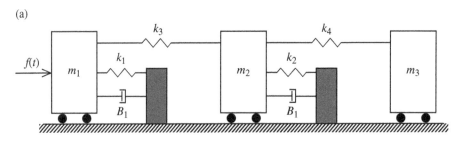

(b)

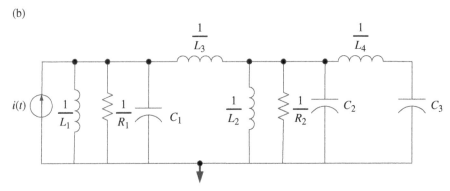

FIGURE 5.28   Mechanical systems and its equivalent electrical circuit model.

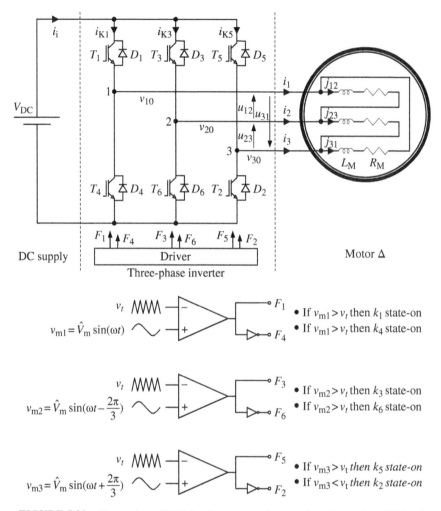

FIGURE 5.29 Three-phase SPWM voltage source inverter for a three-phase RL load.

variable three-phase voltage references on the pulse width modulator circuits:

    i. Develop a fully operational simulation in Simulink.

    ii. Develop a fully operational simulation in PSIM.

    iii. Develop a fully operational simulation in SimPowerSystems Toolbox

5. Develop a simple battery model that contains internal losses, leakage, and energy capacity meter (SOC—state of charge), and design a bidirectional DC/DC converter capable to charge and discharge the battery within its normal range of operation. Develop a photovoltaic model that will serve as input energy provider for

charging the battery. Run simulation studies in MATLAB, Simulink, PSIM, and PowerSystems Toolbox. Design a controller that impose a current charging and discharging profile with commands that will make the system capable to be used as energy storage for a solar energy system.

REFERENCES

[1] EVELEIGH, V.W., Introduction to Control Systems Design, McGraw-Hill, New York, 1972.

[2] OGATA, K., Modern Control Engineering, Prentice Hall, New York, ISBN-13: 978–0136156734, ISBN-10: 0136156738, 2009.

[3] SLOTINE, J.J.E. and LI, W., Applied Nonlinear Control, Prentice-Hall, Englewood Cliffs, ISBN-13: 978–0130408907, ISBN-10: 0130408905, 1991.

[4] MOHAN, N., Power Electronics: A First Course, John Wiley & Sons, Inc., Hoboken, ISBN: 978-1-118-07480-0, 2012.

[5] HART, D.W., Power Electronics, McGraw-Hill, Boston, ISBN: 978-0-07-338067-4, 2010.

[6] LI X., PARK, J.G. and SHIN, H.B., "Comparison and evaluation of anti-windup PI controllers", Journal of Power Electronics, vol. 11, no. 1, pp. 45–50, 2011.

FURTHER READING

CHAKRABORTY, S., SIMÕES, M.G. and KRAMER, W.E., Power Electronics for Renewable and Distributed Energy Systems, 1st edition, Springer, London, 2013.

HATZIARGYRIOU, N., Microgrids: Architectures and Control, 1st edition, Wiley-IEEE Press, Chichester, 2014.

REZNIK, A., SIMÕES, M.G., AL-DURRA, A. and MUYEEN, S.M., "LCL filter design and performance analysis for grid interconnected systems", IEEE Transaction on Industry Applications, vol. 50, no. 2, pp. 1225–1232, 2014.

SIMÕES, M.G. and FARRET, F.A., Modeling and Analysis with Induction Generators, 3rd edition, Taylor & Francis/CRC Press, Boca Raton, 2014.

SIMÕES, M.G., BOSE, B.K. and SPIEGEL, R.J., "Fuzzy logic based intelligent control of a variable speed cage machine wind generation system", IEEE Transactions on Power Electronics, vol. 12, pp. 87–95, 1997.

SIMÕES, M.G., PALLE, B., CHAKRABORTY, S. and URIARTE, C., Electrical Model Development and Validation for Distributed Resources, National Renewable Energy Laboratory, Golden, 2007.

SIMÕES, M.G., ROCHE, R., KYRIAKIDES, E., SURYANARAYANAN, S., BLUNIER, B., MCBEE, K., NGUYEN, P., RIBEIRO, P. and MIRAOUI, A., "A Comparison of smart grid technologies and progresses in Europe and the US", IEEE Transactions on Industry Applications, vol. 48, no. 4, pp. 1154–1162, 2012.

SIMÕES, M.G., MULJADI, E., SINGH, M. and GEVORGIAN, V., "Measurement-based performance analysis of wind energy systems", IEEE Instrumentation and Measurement Magazine, vol. 17, no. 2, pp. 15–20, 2014.

6

INSTRUMENTATION AND CONTROL INTERFACES FOR ENERGY SYSTEMS AND POWER ELECTRONICS

6.1 INTRODUCTION

This chapter requires a broader understanding of electronics and operational amplifiers. However, for reader not very familiar with these topics, it is still possible for them to understand in the following sections how can electronics be used in conditioning signals for power, control, and instrumentation systems.

As a starting point, it should be noted that operational amplifiers serve mostly to condition the electrical properties of measured signals (voltage, current, impedance, or power) in order to make them useful in monitoring and driving machinery and equipment. Such electrical signals would hardly be useful for functions, such as actuation displays, signal processing blocks, or electric drives, without a proper analog signal processing provided by operational amplifiers. Sensors and instrumentation features like differential inputs, feedback, and cancellation of common errors in signal acquisition make these amplifiers powerful and suitable for interfaces between electronic circuits or processes being monitored and computers or for a human–machine interface (HMI). This chapter shows that differential amplifiers are very important for distortion and noise cancellation of the electromagnetic noise induced in conductors and cables of instrumentation as common and/or differential signals, in addition to other analog circuit techniques [1–3].

This book provides the reader a comprehensive understanding on how to model, analyze, simulate, and design circuits, electronics and controls related to power

Modeling Power Electronics and Interfacing Energy Conversion Systems, First Edition.
M. Godoy Simões and Felix A. Farret.
© 2017 John Wiley & Sons, Inc. Published 2017 by John Wiley & Sons, Inc.

systems, power electronics, power quality, and renewable energy. Therefore, it is important to define a simulation platform for such electronics and instrumentation interfaces. PSIM can be useful for evaluating several analog-based circuits, but their models might be ideal or general. Simplorer is another simulation platform that can be used. NI Multisim is an excellent electronic design simulation software, and Matlab has SimElectronics, that is, libraries for modeling and simulating electronic and mechatronic systems. SimElectronics is part of the Simulink Physical Modeling family. Models using SimElectronics are essentially Simscape™ block diagrams. To build up a system-level model with electrical blocks, use a combination of SimElectronics blocks and other Simscape and Simulink blocks. You can connect SimElectronics blocks directly to Simscape blocks. You can connect Simulink blocks through the Simulink-PS Converter and PS-Simulink Converter blocks from the Simscape Utilities library. These blocks convert electrical signals to and from Simulink mathematical signals.

SimElectronics models can be used to develop control algorithms in electronic and mechatronic systems, including vehicle body electronics, aircraft servomechanisms, and audio power amplifiers. The semiconductor models include nonlinear and dynamic temperature effects, enabling you to select components in amplifiers, analog-to-digital converters, phase-locked loops, and other circuits. You can parameterize your models using Matlab variables and expressions. You can add mechanical, hydraulic, pneumatic, and other components to a model using Simscape and test them in a single simulation environment. You can deploy models to other simulation environments, including hardware-in-the-loop (HIL) systems; SimElectronics supports C-code generation.

This chapter presents the most relevant aspects of basic electronic amplifiers related to applications that could affect the quality of operational amplifiers used in signal processing, electronics, and power systems in general. These circuits can be simulated in any electronic-based simulator that fits the reader needs.

6.1.1   Sensors and Transducers for Power Systems Data Acquisition

A data acquisition system aims at making data for automatic controls or intelligible to human decision making. The instrumentation chain is formed by four stages: the physical phenomena, the interface to process measurements (read and write), and the interpretation or processing of a solution and an output, as indicated by Figure 6.1.

Transducers (or sensors) are used in accordance to their specific application suitable for the physical variable being measured and the instrument for data acquisition of such signal. Many data acquisition systems involve more than one sensor (e.g.,

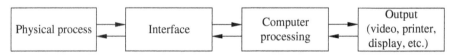

FIGURE 6.1   Basic elements of an instrumentation chain.

a wattmeter). Sensors can be of various forms such as mechanical, optical, or electrical. The electrical sensors can be active or passive. The sensor is passive when the interpretation of the measured data requires an external power source, for example, a resistive divider. On the other hand, when the physical phenomenon being measured induces, alters, or creates a source of internal voltage or current in the transducer, it is called an active one.

6.2   PASSIVE ELECTRICAL SENSORS

The passive electrical sensors can be resistive, capacitive, or inductive. They do not require any external source of energy [4–6].

6.2.1   Resistive Sensors

Equation 6.1 shows a typical resistive sensor formulation:

$$R = \rho \frac{d}{A} \tag{6.1}$$

where

   ρ is the electrical resistivity of the sensor (Ω-m)
   d is the physical length of the element (m)
   A is the cross-sectional area (m^2)

The right-hand side of Equation 6.1 shows the most common physical effects that modify a typical wire resistance (ρ, d, and A), which can be affected by temperature, moisture, compression/traction, degradation, polarization by electric or magnetic field, light, and radiation. In general, the effect of temperature on the electrical resistance of materials can be equated as follows:

$$R_t = R_0 \left(1 + \alpha \Delta t\right) \tag{6.2}$$

where

   R_t is the electrical resistance at a temperature t ($^\circ$C)
   R_0 is the electrical resistance at usual zero-degree Celsius reference temperature ($^\circ$C)
   α is the temperature coefficient
   Δt is the temperature difference with respect to the reference value

Platinum is the most recommendable thermoresistive sensor due its accuracy and stability. It consists generally of a 100Ω resistor at 0°C, and its temperature variation is typically less than 0.004Ω/$^\circ$C. It is widely used in resistance thermometers. As platinum is very expensive, its use is reserved for applications where stability and reading accuracy are essential. In most practical cases, nickel or copper is used instead of platinum.

Some other materials have a higher-order nonlinearity effect mostly due to temperature variations, which can be expressed as follows:

$$R_t = R_0\left(1 + \alpha\Delta t + \beta\Delta t^2 + \gamma\Delta t^3\right) \quad \text{for} \quad \alpha > \beta > \gamma \tag{6.3}$$

Most fluids have a relatively low electrical resistance, and if a fluid surrounds the sensor, its resistance will be affected. For example, in order to have a good electrical insulation of a temperature measurement in a fluid, the sensor resistance wire is wound in a ceramic or epoxy case and covered by a protective tube in order to put it in contact with the medium to measure the temperature. For this reason, the measured response time is very slow (of the order of several seconds or even minutes).

A type of semiconductor, known as thermistor, is a good temperature sensor for its high sensitivity. Furthermore, the temperature coefficient of thermistors can be either positive or negative (PTC or NTC). The variation in resistance of the sensor element as a function of temperature is nonlinear and is approximately −4%/°C. If an NTC is placed under potential differences and undergo heating, this can be used as a flowmeter.

Potentiometers can be used as sensing elements, for example, to determine the angular position or linear displacement of shaft bars. Figure 6.2 illustrates the relationship between the potentiometer resistance and its linear or angular position of the cursor. The output voltage, v_o, is related to the input voltage, v_i, by

$$v_o = v_i x \frac{1}{\dfrac{R_{\text{pot.}}}{R_L} x(1-x) + 1} \tag{6.4}$$

where

$$x = \frac{R_2}{R_1 + R_2} = \frac{R_2}{R_{\text{pot.}}} = \frac{V_o}{V_i}$$

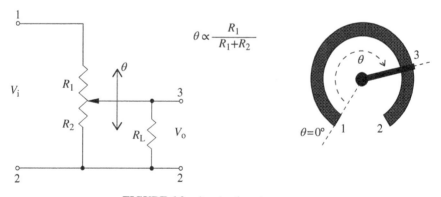

FIGURE 6.2   A potentiometer as sensor.

The resistance of potentiometers varies from $20\,\Omega$ to $200\,k\Omega$ for wire and from $500\,\Omega$ to $80\,k\Omega$ for conductive plastic. This type of sensor introduces a nonlinearity error and, in some cases, some surface contact discontinuity in the variation. The conductive plastic, although without resolution problems, has higher α than metal. The nonlinearity accuracy of the potentiometric sensor varies from 0.1 to 1% for wire wrapped with diameters between 0.5 and 1.5 mm.

The light-dependent resistor (LDR) is a conductive cell whose resistance varies with lighting. LDR is made of cadmium sulfide and responds to some spectrum colors, similar to the human eye response. The resistive sensor is simple, light, and small but needs to carry a current that measures and uses energy for the measurement process.

6.2.2 Capacitive Sensors

As a reference for the understanding capacitive physical sensors, the planar capacitor is used throughout this text, which generally follows the law:

$$C = \varepsilon \frac{A}{d} \tag{6.5}$$

where

$\varepsilon = \varepsilon_r \varepsilon_0$ is the electrical permittivity (F/m)

ε_r is the permittivity of the material relative to the air

$\varepsilon_0 \approx \varepsilon_{ar} = 8854 \times 10^{-12}$ for a dry and clean air (F/m)

A is the physical area of the electrical field over the plate (m^2)

d is the average distance between the plates (m)

The most common application of the capacitive sensor is to measure pressure in some keyboards (computer, mobile phone, and HMI). The separation between plates is as shown in Figure 6.3. In this case, Equation 6.5 becomes

$$C = \varepsilon \frac{A}{d + x}$$

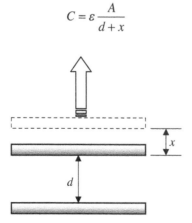

FIGURE 6.3 Capacitive displacement sensor.

where x is the displacement between plates.

The capacitive displacement can be also a device that depends on the insertion of a dielectric material, as portrayed in Figure 6.4:

$$C = \varepsilon_1 \frac{wx}{d} + \varepsilon_2 \frac{w(L-x)}{d} = \frac{w}{d}\left[\varepsilon_2 L - (\varepsilon_2 - \varepsilon_1)x\right]$$

where

x is the displacement in between plates in the outward direction

w is the plate width

L length of the plates

ε_1 and ε_2 are the dielectric constants of the respective materials separating the plates

Another way to vary the capacitance is by positioning rotary axis or tuning tank circuits. Figure 6.5 illustrates such a rotational displacement. Capacitors like this

$$A_R = \int_0^\theta \frac{R^2}{2}\, d\theta \quad \text{and} \quad A_r = \int_0^\theta \frac{r^2}{2}\, d\theta$$

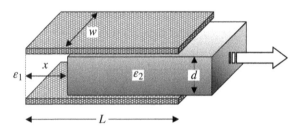

FIGURE 6.4   Insertion capacitive sensor.

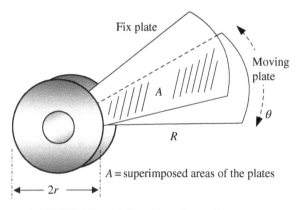

FIGURE 6.5   Rotational-based capacitive sensor.

were very common in the old analog radios. The sector areas with radius R and r, respectively, can be obtained from equations $A_R = \int_0^\theta \left(R^2/2\right)d\theta$ and $A_r = \int_0^\theta \left(r^2/2\right)d\theta$.

The level measurement in liquid systems can use the cylindrical sensor shown in Figure 6.6 where a solid conductive cylinder is concentrically inserted in a conductive cylindrical shell. Both are introduced in a vase containing the liquid whose level is to be measured. The device operates as being two concentric cylinder capacitors. The capacitance per individual coaxial cylinder tube with radius a and b is given by

$$C = \frac{2\pi\varepsilon_r\varepsilon_0}{\ln\left(\dfrac{b}{a}\right)}h$$

Equation 6.6 combines the two capacitances (one immersed in the liquid and the other one above it) to describe this phenomenon:

$$C = \frac{2\pi\varepsilon_0}{\ln(b/a)}\left[L - \left(\varepsilon_r - 1\right)h\right] \tag{6.6}$$

The capacitive sensor is simple, adaptable to several types of measuring environments, and depends on a voltage supply for the measurements (instead of a current). Therefore, it has low consumption and is recommended for battery-powered data acquisition systems.

6.2.3 Inductive Sensors

The most common applications of inductive sensors refer to variations in the magnetic path represented by a magnetic reluctance $\mathfrak{R}$, variations related to the permeability of a magnetic core L, as well as variations in induced voltages. The magnetic permeability ranges from 1.0 (vacuum or air permeability) to thousands (ferrite and other magnetic materials). Magnetic reluctance is analogous to electrical

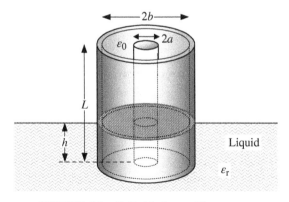

FIGURE 6.6 Cylindrical capacitive sensor.

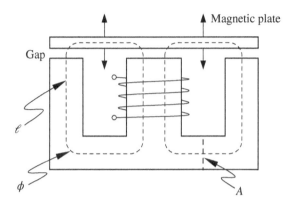

FIGURE 6.7 Reluctance sensor.

resistance, while the MMF is analog to voltage (EMF) and ϕ is analogous to electrical current. The average reluctance of a magnetic core can be given by Equation 6.7, where ℓ is the average path length of the magnetic flux:

$$\Re = \frac{fmm}{\phi} = \frac{Ni}{BA} = \frac{H\ell}{\mu HA} = \frac{\ell}{\mu A} \tag{6.7}$$

An example of a magnetic sensor using variable reluctance is shown in Figure 6.7, where the position of the horizontal plate alters the magnetic reluctance path of the E and I magnetic cores. There is a coil with N turns, and it is assumed that the magnetic reluctance is practically dependent only on the air gap.

As a reference for the physical understanding of inductive sensors, it used the model of an infinite cylinder with uniformly wounded wire windings around its surface. The infinite cylindrical inductor generally follows Equation 6.8:

$$L' = \mu\pi n^2 R^2 \tag{6.8}$$

where

L' is the inductance per meter of coil length
$\mu = \mu_0\mu_r$ is the magnetic permeability of the material
$\mu_0 = 4\pi \cdot 10^{-7}\,H/m$ is the magnetic permeability of the vacuum
μ_r is the relative permeability of the environment inside the cylinder
n is the turn number per meter of length
R is the cylinder radius (m)

As shown by Equations 6.14, 6.15, and 6.8, the electric passive sensors follow similar laws. Their effects are proportional to the component area, inversely proportional to the length of the sensor, and dependent on the material properties of what those passive devices are made.

6.3   ELECTRONIC INTERFACE FOR COMPUTATIONAL DATA IN POWER SYSTEMS AND INSTRUMENTATION

In nature, there are no instantly discontinuous signals; those are in the realm of mathematics only. Industrial processes have a wide variety of signals such as speed, pressure, temperature, current, and a diversity of others that must be converted into digital signals allowing the interpretation by humans or ordinary computers, which by their turn convert virtually any measuring signal from suitable instrumentation sensors or drive external processes.

In principle, the interpretation and processing of all digital computers are done by voltage signals otherwise, the dissipated power (I^2R) would be unacceptable. However, there are some specific cases where a current must be observed and a current–voltage electronic conversion can be used. Therefore, mostly digital voltage computer interfaces are used for connecting the real-world variables, and the external circuits may receive either a voltage or a current command.

In most interface circuits there is one very important device, the operational amplifier. Moreover, any computer data processing would be useless without communication with the outside world through the classic peripherals, such as video monitor, printer, keyboard, and mouse. Interface signals making compatible analog and digital circuits very often need operational amplifiers [7–10].

6.3.1   Operational Amplifiers

The operational amplifier is that one commercially available and ready for use, needing for its operation just an input signal, a device to be activated across its output and a power supply. The user of this circuit does not have to engage in the usual design trade-offs such as gain variations by temperature and humidity, alterations by component aging, radiation effects, and differences between the discrete components making part of them.

An operational amplifier usually has a parallel voltage negative feedback. It is designed to perform various mathematical functions, such as integration, differentiation, summation, signal inversion, scaling, phase shift, and many other functions. In addition, they are designed to meet particular ranges of specifications such as high frequency response and high-gain and low-noise amplification of process signals [1, 3, 8].

6.4   ANALOG AMPLIFIERS FOR DATA ACQUISITION AND POWER SYSTEM DRIVING

Typical operational amplifiers are made of a differential amplifier with two input signals: one is called "inverting," and another is called "noninverting" with a very high gain at the output terminal. This configuration has the property to cancel common signals, that is, similar signals in both inputs, from drivers connected to external environment, caused by electric and magnetic noises (radiated effects) or ripples of power supplies (conducted effects). The following sections illustrate some popular analog processing systems. They are characterized by their high-speed response in detecting signals from natural or industrial processes, such as current, voltage, frequency, capture of zero voltage crossovers, high-slew changes, and others.

6.4.1 Level Detector or Comparator

A level detector serves to sense whether or not a process has reached a desired level, for example, sensing the level in water reservoir or defining the zero-crossing instants of alternating voltages (change from positive voltage to negative and vice versa). Examples of zero-crossing detector and level comparator are shown in Figure 6.8. Whenever it is necessary to limit the output level of the feedback, it used Zener diodes. Therefore, the polarity of the output voltage might have different positive or negative output voltages, defined by the Zener diodes.

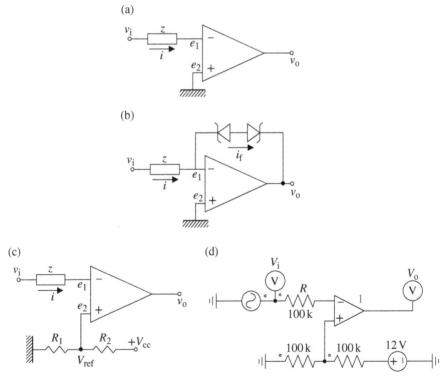

FIGURE 6.8 Level detector circuits for voltages equal and different from zero. (a) Detector of voltage zero crossing, (b) clipped detector of voltage zero crossing, (c) original circuit, and (d) PSIM version.

6.4.2 Standard Differential Amplifier for Instrumentation and Control

There are many ways to design a good quality differential amplifier. It seems to be a usual practice that the amplifier configuration indicated in Figure 6.9 is the most appropriate. Neglecting the excitation currents of the two input amplifiers, the current flowing through resistors R_1, R_2, and R_3 is almost the same and equal to i_1. Moreover, v_{0_1} and v_{0_2} are, respectively, the input signals of the basic differential amplifier. Therefore, these voltage signals can be used to determine the total gain from the individual gains of the two amplifying stages.

The gain of the first stage may be given by Equation 6.9:

$$A_1 = \frac{v_{0_1} - v_{0_2}}{v_1 - v_2} \tag{6.9}$$

The common gain of this amplifier for $v_1 = v_2$ will make $i_1 \cong 0$ and then $v_{0_1} \cong v_1$ and $v_{0_2} \cong v_2$:

$$A_c = \frac{v_{0_1} - v_{0_2}}{v_1 - v_2} = 1 \tag{6.10}$$

The differential gain (for $v_1 = -v_2$) can be obtained from

$$i_1 = \frac{v_{0_1} - v_{0_2}}{R_1 + R_2 + R_3} = \frac{v_1 - v_2}{R_1}$$

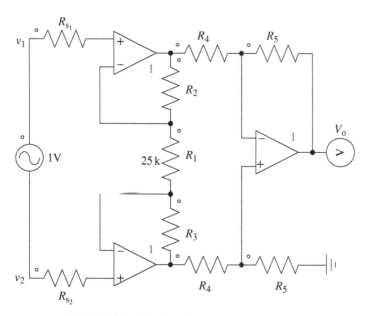

FIGURE 6.9   Standard instrumentation amplifier.

and after rearranging results in

$$A_d = \frac{v_{o_1} - v_{o_2}}{v_1 - v_2} = \frac{R_1 + R_2 + R_3}{R_1} = 1 + \frac{R_2 + R_3}{R_1} \tag{6.11}$$

A merit figure of operational amplifiers is the common-mode rejection ratio (CMMR), which is the relationship between the common and differential gains A_d/A_c established by Equations 6.10 and 6.11. Because the input stage common gain is unitary, then by definition of CMMR comes

$$CMMR = 1 + \frac{R_2 + R_3}{R_1} \tag{6.12}$$

The gain of the classic basic differential amplifier in the second stage is as follows:

$$A_2 = \frac{v_o}{v_{o_2} - v_{o_1}} = \frac{R_5}{R_4} \tag{6.13}$$

Combining the gains of the first and second stages for $R_2 = R_3$ to avoid what is so-called "mirror effect," that is, multiplying Equations 6.4 and 6.6 by each other and simplifying, comes Equation 6.7:

$$A = \frac{v_o}{v_2 - v_1} = \left(1 + \frac{2R_2}{R_1}\right)\frac{R_5}{R_4} \tag{6.14}$$

As can be inferred from Equation 6.14, the input stage can provide a high differential gain by controlling R_1 only and a common-mode unity gain without any matched resistors. Thus, the differential output signal $v_{o_2} - v_{o_1}$ represents a substantial reduction of the common-mode signal and is used to drive a conventional differential amplifier (the second stage), compensating so for any residual common-mode signal. Commercial examples are the LH0036 and AD522 and the precision op-amp-type 3630. The so-called differential instrumentation amplifier 725 is a good quality conventional amplifier and should not be mistaken with an actual instrumentation amplifier.

6.4.3 Optically Isolated Amplifier

In many industrial applications, there is a need to electrically isolate the operator from the process, or the patient from the medical instrumentation, or the power circuit from the controller circuit. Such isolation can be optical, magnetic, or electrical. In the particular case of optical isolation, it used two photodiode–phototransistor pairs, known generically as photocoupler. Figure 6.10 shows a reference pair that compensates for nonlinearity of optically coupled devices. We can note two different grounding points, that is, different grounds for different power supplies.

Commercial optical isolation amplifiers are made with integrated circuits. Therefore, they can achieve a very good match between their photodiode and phototransistor pairs plus their required resistors. In order to represent the coupled values

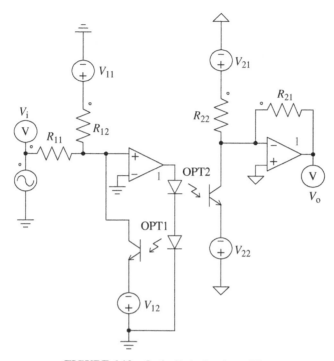

FIGURE 6.10   Optically isolated amplifier.

of the output and input stages in Figure 6.10, the output stage variables are differentiated by number "1" relative to the input stage components and with number "2" for the output stage. Notice in this diagram that in PSIM the number "1" close to the operational amplifier symbol means that this component is a nonideal amplifier (level 1 op-amp), that is, the user can change several of model parameters.

Note that in the circuit of Figure 6.10 the input stage acts as a noninverting amplifier. The impedance of the optically isolated feedback amplifier is exerted by two diodes in series whose cathode is connected to the anode of the other one and this in turn to the ground point. The reference optotransistor connection is optically coupled to the inverting input of the output amplifier through the measurement optotransistor. The gain of this stage is unity, because the biased diodes represent a short circuit across the amplifier output.

A simple way to show that there is a nonlinearity compensation is to establish the following relationships for the input and output stages:

$$I = I_1 - I_2 \tag{6.15}$$

where

$I = I'$
$I_1 = I_1'$ (matched photocoupler made by circuit integration)
$I_2 = I_2' = \dfrac{V_{cc}}{R_2}$

If the power supplies of the input stages and the output-isolating amplifier are coupled, one can say with a good degree of approximation that the approximation $I_1 \cong I_1'$ and $I_2 \cong I_2'$ are quite close and Equation 6.15 is valid. Therefore, the voltage gain can be given from the following:

$$I' = \frac{v_o}{R_1'} = \frac{v_1 - v_2}{R_1} = I \quad \text{or} \quad A_v = \frac{v_o}{v_1 - v_2} = \frac{R_1'}{R_1} \tag{6.16}$$

In practice, the gain is never much greater than unity to ensure that the values of the quiescent photocouplers operate close to one another and an improved linearity is ensured in addition to minimizing electrical fields on them during distinct measurement levels.

The main advantages of optical isolators are (i) less influence to electromagnetic noise, (ii) lower weight and volume, and (iii) higher-speed response. Commercial examples are 3650 and the 3652 from Burr-Brown. Such amplifiers exhibit a nonlinearity of 0.05%, with a common-mode voltage isolation of 2 kV, CMRR higher than 60 dB at 140 Hz, and a 15 kHz band pass. Usually optocouplers are required to isolate the ground of a microcontroller, or a DSP, or a computer in general from transistors switching in a three-phase or in a single-phase inverter.

6.4.4   The V–I Converter of a Single Input and Floating Load

A direct action of a computer output commanding an electric current-dependent process, such as a relay coil, will need a voltage-to-current converter. Figure 6.11 shows a single-input V–I converter with a floating load. The conversion between the load current and the input voltage is established as

$$I_L \cong \frac{1}{R_f} v_i \tag{6.17}$$

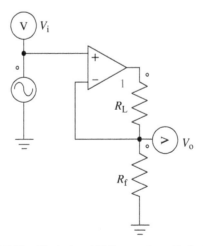

FIGURE 6.11   Single-input V–I converter with floating load.

6.4.5 Schmitt Trigger Comparator

In every electrical process, there is always some noise or ringing at load switching instants, which may cause problems with level detectors or microprocessors, chattering them from one state to the other. One way to protect from such ringing is to use a hysteresis band around the level to be compared, such that the closing voltage signal happens at a higher level with respect to the opening voltage signal. The signal at lower voltage level will determine the signal capture in a lower level. Figure 6.12 illustrates this example.

6.4.6 Voltage-Controlled Oscillator (VCO)

The voltage-controlled oscillator or voltage–frequency converter (VFC) can be used as a signal modulator or in pulse width modulation (PWM) or even to measure voltage levels with a frequency output. Figure 6.13 shows that the control voltage V_c is converted into triangular or rectangular output pulses (v_{o_2} and v_{o_1}). Resistors R_1 and R_2 only serve to lose the firm voltage sources (small internal resistances) across the inverting and noninverting input terminals (virtual input short circuit) of the operational amplifier. Resistor R_4 defines the duty cycle modulation. Commercial examples of this circuit are the 3130 (FET) and 3160 coupled to an external BJT buffer.

6.4.7 Phase Shifting

The phase-shift amplifier can impose a prescribed phase shift between input and output signals while maintaining the same amplitude (Figure 6.14). Equations 6.18 show the input of the inverting and noninverting inputs of the operational amplifier:

$$e_1 = v_i - i_1 R_1 = v_i - \frac{v_i - v_o}{R_1 + R_2} R_1$$

$$e_2 = \frac{v_i R}{R - j\dfrac{1}{\omega C}}$$

$$(6.18)$$

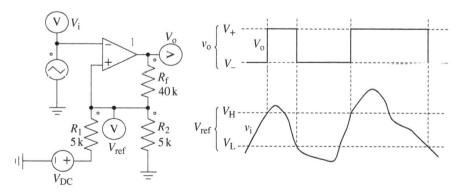

FIGURE 6.12 Schmitt trigger comparator.

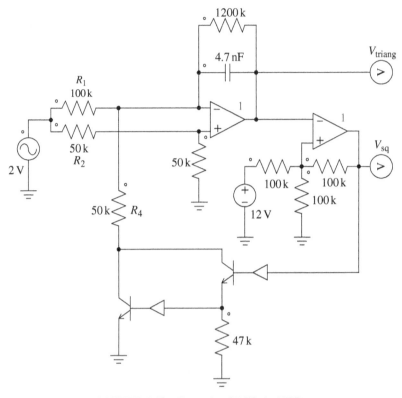

FIGURE 6.13 Example of VCO (or VFC).

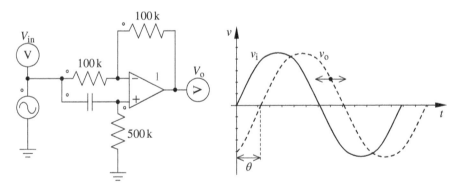

FIGURE 6.14 Phase-shift amplifier.

Equating e_1 and e_2 by considering the virtual ground connection for the inverting and noninverting inputs of the operational amplifier,

$$v_i - \frac{v_i - v_o}{R_1 + R_2} R_1 = \frac{v_i R}{R - j\dfrac{1}{\omega C}}$$

or

$$v_i \left(1 - \frac{R_1}{R_1 + R_2} - \frac{R}{R - j\frac{1}{\omega C}} \right) = -v_o \frac{R_1}{R_1 + R_2}$$

Simplifying this expression for $R_1 = R_2$ results in

$$\frac{v_o}{v_i} = \frac{\omega CR + j}{\omega CR - j} \tag{6.19}$$

If Equation 6.19 is converted to its polar form, it becomes

$$\left| \frac{v_o}{v_i} \right| \underline{|\theta_1|} = |1| \, 2 \tan^{-1} \left(\frac{1}{\omega CR} \right) \tag{6.20}$$

Equation 6.20 ensures the unity gain amplitude. It is assumed that the phase between input and output signals is dependent only on R and C for a given frequency ω. Therefore, by opting for a variable R rather than a variable C, the adjustment for the resistance can be established by the following conditions:

1. If $R = 0$ (short circuit),

$$\frac{v_o}{v_i} = 1 \underline{|180°}$$

2. If $R \rightarrow \infty$ (open circuit),

$$\frac{v_o}{v_i} = 1 \underline{|0°}$$

In conclusion, one can get a phase variation between the output and input signals by a simple R variation from zero (short circuit) to infinity (open circuit). Similarly, in the circuit of Figure 6.14, if the capacitor is interchanged with the resistor,

$$\left| \frac{v_o}{v_i} \right| \underline{|\theta_1|} = |1| \, -2 \tan^{-1} \left(\frac{1}{\omega CR} \right) = |1| \underline{|-\theta}$$

In this case, for $R = 0$,

$$\frac{v_o}{v_i} = 1 \underline{|-180°}$$

Both aforementioned conditions allow a variation phase from 0° to 360° between the input and output signals while maintaining the same amplitude. In practice, the short-circuit and open-circuit conditions will cause a small amplitude variation (the operational amplifier input forces an actual short circuit between the input signals).

6.4.8   Precision Diode, Precision Rectifier, and the Absolute Value Amplifier

Precision diodes are circuits that do not have the threshold voltage (diode voltage drop) and use an operational amplifier with proper feedback. Figure 6.15 shows a circuit that for any positive half cycle of the input signal, the output has that signal, because of the virtual short circuit of the input and output feedback. There is no gain. The output receives a voltage from the source as if the diode is closed (ON) between input and output. For the negative half cycle of the input signal, the op-amp output is negative, the diode is OFF, and there is no feedback. Therefore, the gain is zero. An adaptation of this circuit works as a peak detector (Figure 6.16). The capacitor charges to the maximum of the input voltage and keeps that voltage. It may be necessary to add a very high resistance in parallel with the capacitor in order to discharge it, but that will depend on the application.

Figure 6.17 illustrates the half-wave rectifier with a voltage gain, while the circuit of Figure 6.18 illustrates the full-wave rectifier or the one that takes the absolute value of the input signal.

A full-wave precision rectifier with fewer components is depicted in Figure 6.19. This version is not as accurate as the circuit indicated in Figure 6.18 since circuit of Figure 6.19 relies only on the accuracy of a resistive divider for the positive half cycle in the operational amplifier output. Its accuracy is affected by the need of high-precision components and the load; otherwise, it distorts the output. The advantages are simplicity and cost.

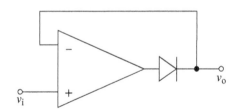

FIGURE 6.15   Precision diode (gain one).

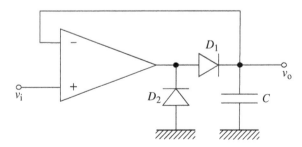

FIGURE 6.16   Peak detector.

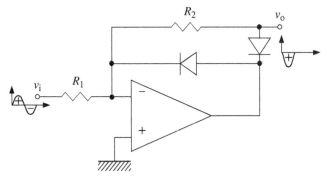

FIGURE 6.17 Half-wave precision rectifier (with nonunitary gain).

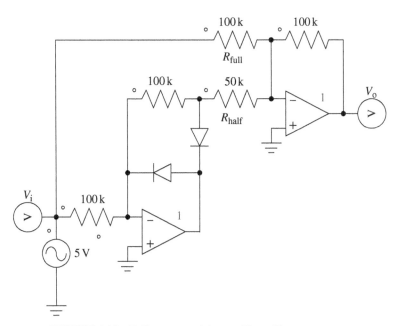

FIGURE 6.18 Full-wave precision rectifier with two op-amps.

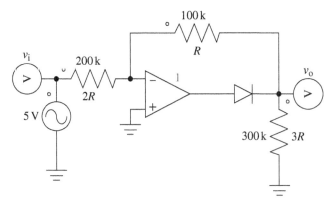

FIGURE 6.19 Full-wave precision rectifier with only one op-amp.

6.4.9 High-Gain Amplifier with Low-Value Resistors

High-gain amplifier requires high-value resistances, in the order of $1\,\text{M}\Omega$ or more. These resistors usually have significant leakage capacitance and inductance, limiting the bandwidth of the amplifier. Therefore, it is necessary to have some kind of compensation for this problem. The classical transformation star-delta supports for the equivalent feedback impedance of Figure 6.20, as depicted in Equation (6.21). Such transformation can be applied to high-gain feedback amplifiers, as indicated in Figure 6.21:

$$Z_f = \frac{Z_1 Z_2 + Z_1 Z_3 + Z_2 Z_3}{Z_2} \tag{6.21}$$

An ordinary configuration of an inverting amplifier with gain 100 and $10\,\text{k}\Omega$ input impedance would need a $R_0 = 1\,\text{M}\Omega$ resistor. Using the equivalent circuit given in Figure 6.20 for the voltage gain and using Equation 6.14 for the values $R_1 = R_3 = 10\ \text{k}\Omega$ results in

$$10^6 = \frac{10^4 R_2 + 10^4 \cdot 10^4 + 10^4 R_2}{R_2}$$

Isolating R_2 in the previous equation and solving it, one gets $R_2 = 1024\ \Omega$.

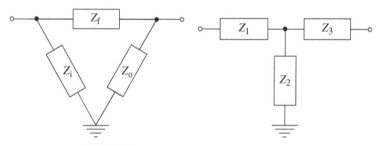

FIGURE 6.20 Star-delta equivalent circuits.

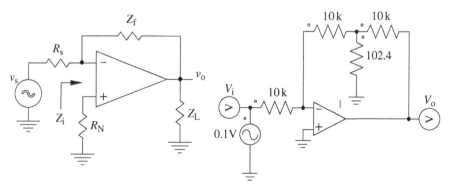

FIGURE 6.21 Star-delta equivalent circuits in operational amplifiers.

6.4.10   Class B Feedback Push–Pull Amplifiers

The class B feedback push–pull amplifiers minimize the distortions caused by the voltage zero crossing ("crossover") of diode junctions. Figure 6.22 illustrates a circuit where the feedback crossover distortion is minimized by using a pair of complementary (NPN and PNP) bipolar junction transistors. Further design details are available in Ref. [11].

6.4.11   Triangular Waveform Generator

A triangular waveform generator is very useful for PWM schemes. Typically a triangle wave is generated either by an integrated square wave or by a current source that switches the polarity of a current charging a perfect capacitor. The triangular/square-wave circuit shown in Figure 6.23 can be used to produce triangular carrier wave for PWM [12, 13].

The parameters of the circuit shown in Figure 6.23 can be easily modified for other amplitudes of triangular waveforms or other operating frequency from [9–12, 14]

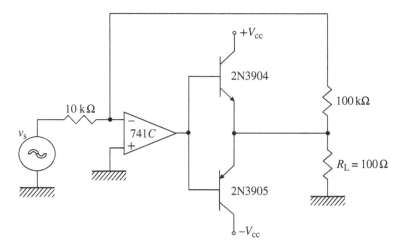

FIGURE 6.22   Class B feedback push–pull amplifier.

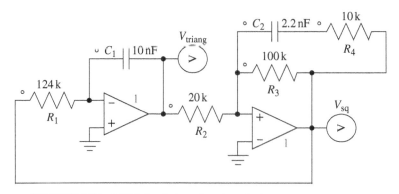

FIGURE 6.23   PSIM simulation of a triangular/square-wave generator.

$$f = \frac{1}{4R_1C_1}\frac{R_2}{R_3}$$

The limits of the square waveform are given by the operational DC sources, and the triangle waveform is expressed by

$$V_{triang} = \pm V_s \frac{R_2}{R_3}$$

where V_s is the DC source amplitude.

The reader must pay attention to the operational amplifier with R_3 feeding back to the noninverting input. That is correct, because this circuit operates as a Schmitt trigger comparator with a hysteresis window around 0.0 V. If the feedback is changed to the inverting input, the same circuit operates as an inverting amplifier, instead of a comparator.

6.4.12 Sinusoidal Pulse Width Modulation (PWM)

Three-phase inverters and single-phase inverters use a symmetrical triangle waveform for sinusoidal pulse width modulation (SPWM), while DC/DC converters usually use the triangular waveform with an offset, such that it starts from 0.0 V swings to the peak and returns to zero. An illustration of the triangular waveform circuit is available in Figure 6.23, depicting the PSIM simulation scheme. Figure 6.24 shows the analog sinusoidal pulse-width modulation as simulated in PSIM, whereas Figure 6.25 shows the triangular waveform constructed from the square-wave generator.

The simplified circuit in Figure 6.24 should pass through an amplifier circuit made for adjusting the offset. However, a circuit so simplified is useless when you want to build it up in a laboratory project. Figure 6.26 shows a good feasible circuit with all details to be used in a laboratory project.

Some spikes may occur in the square-wave output due to the fast discharge transition of the integrating capacitor. In general, they do not cause problems but if they interfere in other electronic devices and a derivative branch may be added to this circuit like the one formed by R_4C_2. The resulting clean wave shape is shown in Figure 6.27.

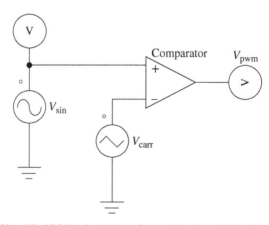

FIGURE 6.24 Simplified PSIM simulation of an analog sinusoidal pulse width modulation.

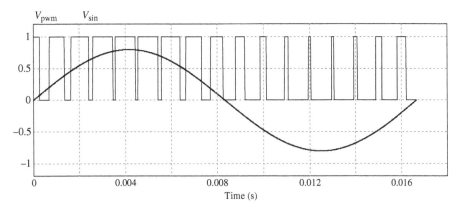

FIGURE 6.25   Triangular waveform of a simplified square-wave generator.

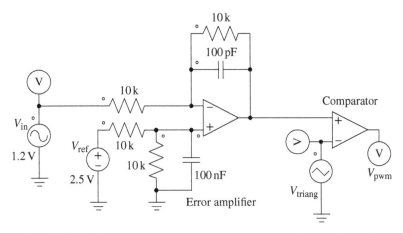

FIGURE 6.26   Symmetrical square-wave generator and their error integration.

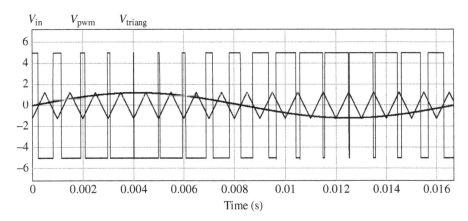

FIGURE 6.27   Output of an analog sinusoidal pulse width modulation (SPWM).

6.5 LABORATORY PROJECT: DESIGN A PWM CONTROLLER WITH ERROR AMPLIFIER

PSIM allows the simulation of PWM controllers and implementation of a sine-wave for power converters in a very simple way, as shown in Figure 6.26. The output of this SPWM converter is presented in Figure 6.27.

In this laboratory project, you should design a PWM controller with an error amplifier as indicated in Figure 6.26 and its triangular wave generator with a 20 kHz carrier frequency, and apply the SPWM for either a single-phase or a three-phase power converter. Would you suggest an analog filter for the sine wave modulation? Minimize the power side components with higher triangular frequencies in such a way to guarantee similar power output quality.

6.6 SUGGESTED PROBLEMS

1. Instead of using a real-life-based SPICE model for an operational amplifier, the model of Figure 6.28 can be used in any circuit simulator. You need a controlled voltage source that depends on two points with infinite resistance. Implement this equivalent model and simulate all the op-amp-based circuits described in this chapter. You should save your equivalent voltage-controlled model as a subsystem with two inputs (inverse and no inverse) with one output (V_o), and the differential gain should be as high as 200 000.

2. Simulate in PSIM the circuit shown in Figure 6.29 for feedback gains of 0.5, 2, and 5 for a linear LED illumination control at $f = 60$ Hz. Assume that the illumination intensity is proportional to the actuator voltage from 0 to 100%. If the amplitude of the input signal was 6V, what would be your recommendation for practical values of $\pm V_{\text{DC}}$ to attend to the compliance limits of the operational amplifier? From these simulation parameters, the actuator voltage is given by

$$V_{\text{act}} = k_p V_\varepsilon + k_p k_i \int dV_\varepsilon dt$$

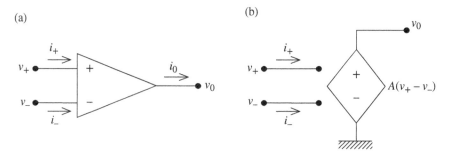

FIGURE 6.28 Equivalent model of an op-amp easy to implement in any circuit simulator: (a) the operational amplifier and (b) the use of a voltage-controlled source with a very high gain of the two input terminals.

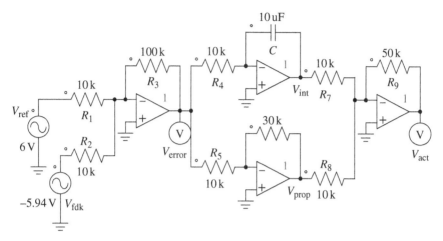

FIGURE 6.29   Proportional–integral amplifier.

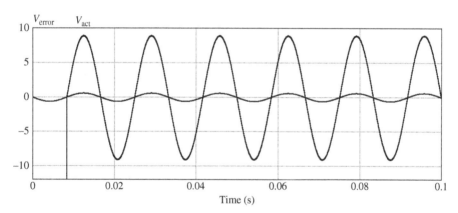

FIGURE 6.30   Error and actuator voltages of a proportional–integral amplifier.

Could you double the proportional and integral gains of this circuit based on the values given in Figure 6.29 and plot as given in Figure 6.30? What would be the limits of stability of such modifications on the proposed control? Please complete the feedback loop to keep a constant illumination level independent of a standard 110 V voltage source under variable conditions of ±10%.

3. What is the minimum error which is expected in the electrical measurement on a circuit that has input resistance of 500 kΩ when it is used to measure a resistive divider with a 250 kΩ load and high levels of stress? Make a circuit sketch that could reduce this error without moving inside the instrument using an operational amplifier?

4. Using a table of typical parameters of an operational amplifier, calculate the error in volts measured at the amplifier output voltage below with and without the resistance in the noninverting input of the two-input adder, depicted in Figure 6.31.

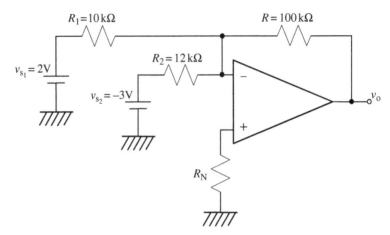

FIGURE 6.31 Two-input adder.

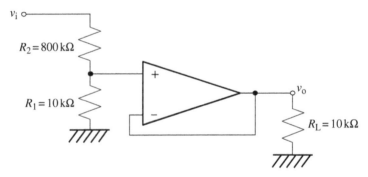

FIGURE 6.32 Voltage divider with a boost amplifier.

5. Design and simulate in Matlab/Simulink a simulation to solve the equation $dy/dx = -10y + 12$. Design an analog computer-based solution, that is, a circuit with operational amplifiers in PSIM, and compare the strategies. Discuss the scaling factors of your circuit simulation as compared to the actual variables of the original differential equation.

6. Calculate the percentage error in the measurement of output voltage of a resistive divider such as in the following figure when not used as "buffer" output (Figure 6.32).

7. Compare the expected percentage error of an instrument based on an amplifier with the following offset values: $e_{os} = -0.2$ mV, $i_1 = 22$ nA, and $i_{os} = 2$ nA. Determine the amplifier error when it uses the configuration given in the figure below with and without the compensation resistance in the noninverting terminal (Figure 6.33).

8. Simulate the output signal of the amplitude variations in operational amplifier with an input signal $v_i = 0.005 \text{sen}\omega t$ for $f = 60$ kHz, changing the input

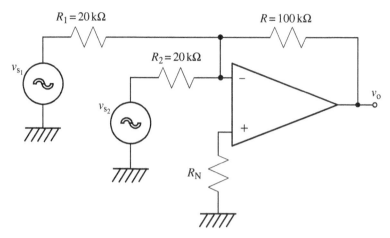

FIGURE 6.33 Inverting two-input adder with offset and drift error compensation.

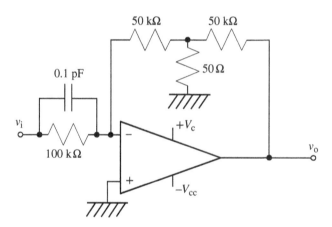

FIGURE 6.34 High-gain amplifier with low-value resistors.

capacitance of the cable as shown below. What is the range if the input signal is DC to the same extent? What is the instrument input impedance for 60 kHz? Verify the simulation for a minimum recommendable value of $\pm V_{DC}$ (Figure 6.34).

9. Calculate and use a PSIM analysis to check the passband of the first-order filter below designed for the input of a digital multimeter. The RC impedance input is $10^4 - j10^6\,\Omega$ for the frequency of 10 kHz, and the amplifier unit gain is 12 MHz. Design your system based on the device gain amplifier product versus their passband (Figure 6.35).

10. A simple PWM-based PI controller with an error amplifier can be built with operational amplifiers. The PWM controller impresses voltage in an RL circuit, that is, a first-order system which can be simulated as a RC low-pass filter. Such system is depicted in Figure 6.35 as an analog computer-based simulation.

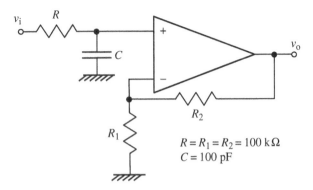

FIGURE 6.35 Amplifier with a first-order filter.

$R = R_1 = R_2 = 100\ \text{k}\Omega$
$C = 100\ \text{pF}$

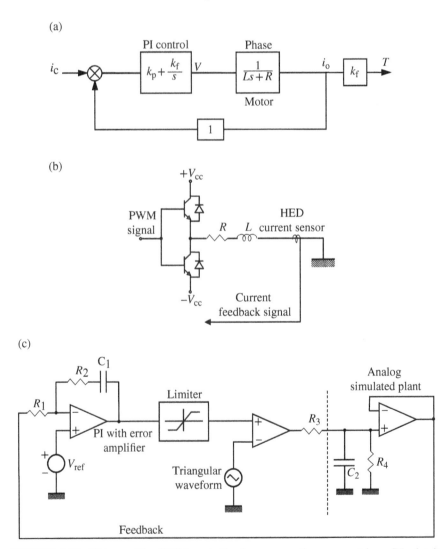

FIGURE 6.36 PI loop with a PWM controller for commanding a current in an RL circuit. (a) Block diagram, (b) power output with push–pull transistors, and (c) PI analog controller with PWM.

Suppose the parameters are $R = 100\,\Omega$, $L = 0.15\,\text{H}$. The operational amplifiers are connected to a $+15\,\text{V}$ and $-15\,\text{V}$ power supply, and the output of such operational amplifier is capable of driving the RL load, or a push–pull transistor output, similar to Figure 6.36, can be implemented in a real laboratory setup. The PWM frequency is 1 kHz, so a triangular waveform can be implemented such as that shown in Figure 6.23.

1. Design the PI control using classic control techniques (root-locus or Bode diagrams) using Matlab.

2. Simulate the closed-loop control in Matlab and Simulink.

3. Simulate the closed-loop control using operational amplifiers as shown in Figure 6.35.

4. Implement a real laboratory prototype, and fine-tune the PI control for best transient response.

5. Write a report about how a model-and-simulation-based design use in this problem can be expanded for real-life large projects.

6. Design a PI loop with a PWM controller for commanding the current in an RL circuit as shown in Figure 6.36: (a) control system block diagram of a PI controller for commanding current in an RL circuit, (b) push–pull transistor and diode circuit with a Hall-effect (HED) current sensor, and (c) analog computer simulation-based study of PI with error amplifier, PWM, and a first-order plant with $R_3 C_2$ noninverting first-order circuit emulating the RL dynamics, where $R_4 \gg R_3$ just to discharge C_2 when the circuit is turned off.

REFERENCES

[1] GRAEMME, J.G., TOBEY, G.E. and HUELSMAN, L.P., Operational Amplifiers: Design and Applications, Burr-Brown Publishers, Tokyo, 473 p., 1971.

[2] SILVEIRA, P.R. and SANTOS, W.E., Discrete Automation and Control, Érica, São Paulo, 229 p., 1998.

[3] DIEFENDERFER, A.J., Principles of Electronic Instrumentation, W.B. Saunders Co. Publisher, Philadelphia, 667 p., 1972.

[4] SLOTINE, J.J.E. and LI, W., Applied Nonlinear Control, Prentice Hall, Englewood Cliffs, 461 p., 1991.

[5] ZELENOVSKY, R. and MENDONÇA, A., PC: A Practical Guide of Hardware and Software, 3rd edition, MZ Publishers, Rio de Janeiro, 760 p., 1999.

[6] REGTIEN, P.P.L., Electronic Instrumentation, VSSD Publisher (Vereniging voor Studie en Studentenbelangen te Delft), Delft, 978-90-71301-43-8, 2005.

[7] BOWRON, P. and STEPHENSON, F.W., Active Filters for Communications and Instrumentation, McGraw-Hill Book Company, Berkshire, 285 p., 1979.

 [8] AUSLANDER, D.M. and AGUES, P., Microprocessors for Measurement and Control, Osborne/McGraw-Hill, Berkeley, 310 p., 1981.

 [9] BOLTON, W., Industrial Control and Measurement, Longman Scientific and Technical, Essex, 203 p., 1991.

[10] EVELEIGH, V.W., Introduction to Control Systems Design, TMH edition, McGraw-Hill, New York, 624 p., 1972.

[11] BOYLESTAD, R. and NASHELSKY, L., Electronic Devices and Circuit Theory, PHB, New York, ISBN-13: 000-0132622262, ISBN-10: 0132622262, 2012.

[12] HOROWITZ, P. and HILL, W., The Art of Electronics, Cambridge University Press, London, 716 p., 2002.

[13] MALVINO, A.P., Electronic Principles, McGraw-Hill Book Company, New York, 2006.

[14] WAIT, J.V., HUELSMAN, L.P. and KORN, G.A., Introduction to Operational Amplifier Theory and Applications, International Student Edition, McGraw-Hill-Kogakusha, Tokyo, 396 p., 1975.

7

MODELING ELECTRICAL MACHINES

7.1 INTRODUCTION TO MODELING ELECTRICAL MACHINES

The study of electrical machines is a classic subject in electrical engineering and usually involves the understanding of transformers and rotating devices. Their models are well understood mostly by students and engineers who eventually work in design and applications of machine, generators, motor drives, and automation. When the electrical machine model is operational, the designer can use it in a more complex open-ended study, such as in a control system, a power system, a power electronic system, or even in a mechatronic or robotic application.

It is always important to understand the best operation when a machine is connected to a load either mechanical (shaft) or electrical (when the machine operates as a generator). For electrical loads, it must be taken into consideration that they are not always a permanent device. They may be variable, may work at lower efficiency points, and may become overloaded or short-circuited in some situations. This whole subject becomes a lot more complex when power electronics and digital controls are also part of the overall system, coming to further complexity when this interacts with a power system feeding power from a substation or when the power becomes bidirectional and is injected into the utility grid.

Several simulation programs can give a good platform to help the electrical engineer to find a reasonable solution with quite accurate loading predictions and

Modeling Power Electronics and Interfacing Energy Conversion Systems, First Edition.
M. Godoy Simões and Felix A. Farret.
© 2017 John Wiley & Sons, Inc. Published 2017 by John Wiley & Sons, Inc.

minimum costs. This chapter gives the foundations of this subject that might be very important for students, engineers, and designers to work through a model based on a simulation design approach.

7.2 EQUIVALENT CIRCUIT OF A LINEAR INDUCTION MACHINE CONNECTED TO THE NETWORK

There are two simple ways of simulating an induction machine (IM) in PSIM: either using an equivalent circuit PSIM block made up with electrical components or using the premade equivalent block that can be embedded in the electrical circuit simulation. This model can be used for the evaluation of three-phase machines, assuming that they are symmetrical and balanced. Therefore, power factor, losses, and the general performance can be established under a per-phase basis. A third possibility is using MATLAB/Simulink block diagram modeling techniques.

Figure 7.1 shows an equivalent per-phase circuit model for a linear IM. Linear modeling is understood as an approximate representation of an underloaded machine (no core saturation). Since the rotor resistance is modified by the slip ratio (s) and the rotor current is not corrected for the right induced frequency (since this model is for steady-state evaluation), it is expected that this model does not have an accurate transient representation of a machine for one phase, but it is very helpful in validating power factor, efficiency, and speed regulation studies.

The equivalent circuit of Figure 7.1 has been simulated with the physical parameters listed in Table 7.1. If the IM rotation is under the synchronous speed, it will act as a motor, and if it is above, it will act as a generator. A detailed mathematical description of this model is presented in Ref. [1]. The equivalent resistance represented by R_{mech} is a nonlinear "resistor" which includes the slip factor s as depicted by Equation 7.1:

$$R_{mech} = R_2 \frac{(1-s)}{s} = \frac{R_2}{\left(\dfrac{n_S}{n} - 1\right)} \qquad (7.1)$$

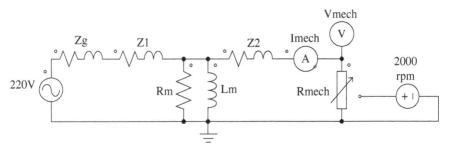

FIGURE 7.1 Per-phase equivalent circuit of an induction machine working as motor/ generator.

Figure 7.1 shows the virtual resistance, R_{mech}, that can be positive or negative, depending if the rotor speed n is above or below the corresponding grid frequency, that is, the machine will be either producing power (generator) or consuming power (motor).

The simulated output voltage of the IM model working as an electrical generator must be at a speed above the synchronous. The voltage across and current through the mechanical equivalent resistance $R_{mech} = R_2(1-s)/s$ are presented in Figure 7.2. The 180° phase shift between output voltage and current shows a negative power (power injection into the network) with respect to the source, since this machine is operating at a speed above the synchronous rotation.

TABLE 7.1 The Induction Machine Parameters

Parameter	Variable	Magnitude
V_{ph}	Rated phase voltage	220 V
I_{ph}	Rated phase current	60 A
n_s	Rated rotor speed	1800 rpm
p	Number of poles	4 poles
R_s	Source resistance	1.53 Ω
L_s	Source inductance	5.74 mH
R_1	Stator resistance	0.622 Ω
R_2	Rotor resistance	0.912 Ω
L_1	Stator inductance	1.14 mH
L_2	Rotor inductance	1.14 mH
R_m	Core losses	114 Ω
L_m	Magnetizing inductance	127 mH

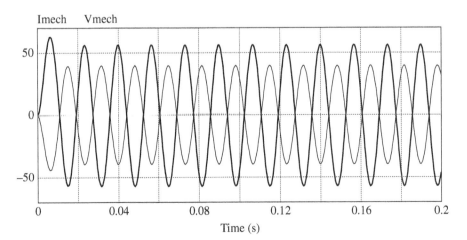

FIGURE 7.2 Simulation results of the IM with parameters listed in Table 7.2.

7.3 PSIM BLOCK OF A LINEAR IM CONNECTED TO THE DISTRIBUTION NETWORK

Another possibility in modeling a linear IM connected to the distribution network in PSIM is presented in Figure 7.3. The parameters are listed in Table 7.2. A brief mathematical description for such model is presented in Refs. [1–3].

For the calculation of torque and power, it was used a mathematical block with the following expressions:

$$P = T\omega \tag{7.2}$$

$$\omega = 2\pi \frac{n}{60} \tag{7.3}$$

Therefore, the mechanical shaft power is calculated by Equation 7.4.

$$P = T\pi \frac{n}{30} \tag{7.4}$$

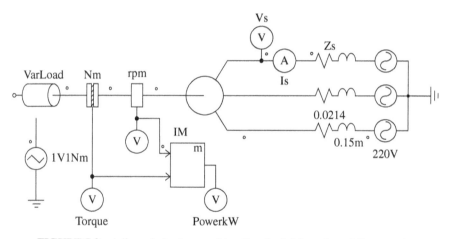

FIGURE 7.3   A linear induction machine directly fed from the public network.

TABLE 7.2   The Saturated Induction Machine Parameters

Parameter	Variable	Magnitude
V_{ph}	Rated phase voltage	220 V
I_{ph}	Rated phase current	42 A
n_s	Rated rotor speed	1740 rpm
p	Number of poles	4 poles
R_s	Source resistance	0.011 Ω
L_s	Source inductance	0.047 mH
R_1	Stator resistance	0.294 Ω
R_2	Rotor resistance	0.156 Ω
L_1	Stator inductance	1.39 mH
L_2	Rotor inductance	0.74 mH

The output voltage of the IM feeding a mechanical load is depicted in Figure 7.4. Notice the initial higher transient current during the start-up of this machine. The duty cycle of the ramp voltage $D = T_{ramp}/T$ begins at T_i, and it should last until the total simulation time T_{end}. The ramp frequency should be long enough to last for the testing period, which is given by $f_{ramp} = 1/T_{end}$.

Figure 7.4 shows the induction machine output voltage connected to the distribution network and the $P \times T \times RPM$ characteristics, which are portrayed in Figure 7.5. Notice that if the torque T_c changes, it also changes the initial transient state time necessary to pass from rest to the steady state. Then, it acts in conjunction with the inertia moment (J) supplying losses, for example, for $T_c = 1\,Nm$ and $J = 0.0005\,kg{\cdot}m^2$. In addition, when doubling resistance R_2, the $P \times T \times RPM$ characteristic tends to be rounded. The oscillations at the beginning of the $P \times T \times RPM$ characteristic are

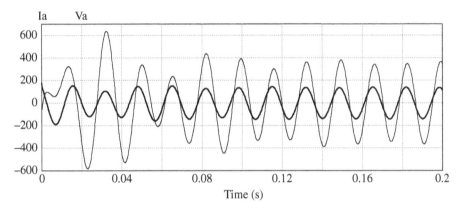

FIGURE 7.4   Linear IM output voltage connected to the distribution network.

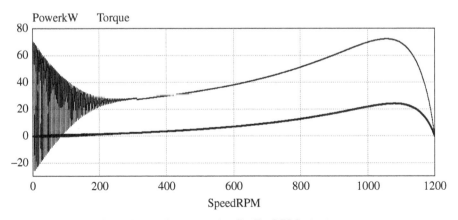

FIGURE 7.5   Characteristics $P \times T \times RPM$ of a linear IM.

expected because the initial conditions for the IM model affect the response (windings initial currents, initial angular speed, and oscillation due mechanical inertia).

7.4   PSIM SATURATED IM MODEL CONNECTED TO THE DISTRIBUTION NETWORK

The model presented in Figure 7.6 can be used with saturated iron properties input by the user. The physical parameters are listed in Table 7.2, and the magnetizing characteristics, that is, terminal phase voltage versus magnetizing phase current, are available in Table 7.3. A brief mathematical description of a model like this is presented in Refs. [1, 4, 5].

The output voltage of the IM simulation is plotted in Figure 7.7 using the magnetizing characteristic given in Table 7.3 when feeding a mechanical load. Notice again the initial higher current during the machine start-up. The duty cycle of the ramp voltage $D = T_{ramp} / T$ begins at T_i, and it should last till the total simulation time T_{end}. The ramp voltage frequency should be long enough to last for the testing period, which is given by $f_{ramp} = 1/T_{end}$.

With same configuration presented in Figure 7.6, the plot of the $P \times T \times RPM$ characteristics is obtained, as shown in Figure 7.8. Notice that if the torque T_c changes it will change the initial inrush current necessary to pass from zero initial conditions state to the steady state. Then, it works in conjunction with the inertia moment to supply losses; for example, $T_c = 1$ Nm and $J = 0.0005$ kg·m². Also, if R_2 is increased, the $P \times T \times RPM$ characteristic has a trend to be rounded. The oscillations at the beginning of the $P \times T \times rpm$ characteristic are expected since the initial transient state of the IM is included (windings initial currents, angular speed, and mechanical inertia).

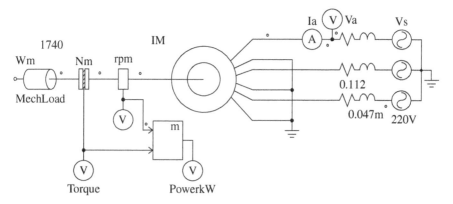

FIGURE 7.6   Saturated induction machine directly fed from the public network.

TABLE 7.3 The Saturated Magnetizing Characteristic of an IM

V_{ph}	I_{ph}
1.241	0.0782
1.969	0.0751
2.519	0.0742
2.963	0.0728
3.371	0.0714
3.742	0.0698
4.104	0.0682
4.474	0.0664
4.813	0.0646
5.178	0.0625
5.606	0.0601
5.897	0.0584
6.257	0.0563
6.612	0.0544
6.975	0.0525
7.641	0.0493
8.292	0.0466

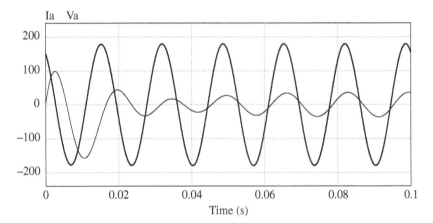

FIGURE 7.7 Output voltage of a saturated IM connected to the distribution network.

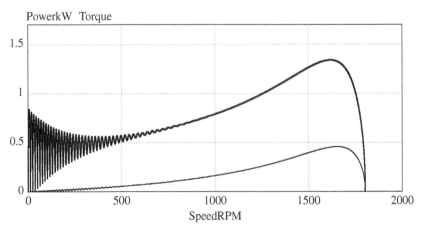

FIGURE 7.8 Characteristics $P \times T \times$ RPM for a saturated IM.

7.5   DOUBLY FED INDUCTION MACHINE CONNECTED TO THE DISTRIBUTION NETWORK

A good model to study the characteristics of a doubly fed induction machine (DFIM) is presented in Figure 7.9 to drive a mechanical load as the one represented in Table 7.4. A mathematical description of this model is presented in Ref. [1]. In the model of Figure 7.9, instead of a power converter, a simplified three-phase generator is connected to the rotor winding and can be easily adjusted in order to obtain different features for a diversity of controlled conditions. The physical parameters used in the generator of Figure 7.9 are listed in Table 7.4. The output voltage can be regulated by power converters with a lower level using, for example, a buck converter or to a higher level with a boost converter or a double-PWM converter at the rotor side with a transformer to match the rotor voltage to the utility grid. The output voltage of the DFIG connected to the distribution network is shown in Figure 7.10. The corresponding $P \times T \times$ RPM characteristics of this machine can also be plotted using the same technique used for Figure 7.7.

If R_2 is doubled, the $T \times$ RPM characteristic tends to be rounded. The oscillation at the beginning of the $T \times$ RPM characteristic is expected since the initial transient state of the IM is included in this simulation (windings initial currents and mechanical inertia). The output power and rotor speed are calculated by [1–3]

$$P_r = T\omega_r$$

$$\omega_r = \pi \times p \times \text{rpm}/60$$

Remember that if the torque T_c changes, it changes the initial transient state time necessary to pass from rest to the steady state. Then, the DFIM acts in conjunction with its inertia moment to supply losses. The load in this simulation varies from 0 to 12 Nm.

A second solution to represent the DFIM connected to the distribution network by using the per-phase equivalent circuit of the IM working as motor/generator is shown in Figure 7.11 whose output voltage shown in Figure 7.12.

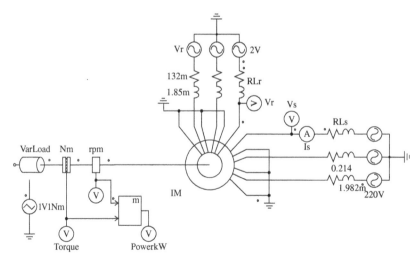

FIGURE 7.9   DFIM connected to the distribution network.

TABLE 7.4   Parameters of the DFIM

Parameter	Magnitude
Rated stator voltage	220 V
Rated stator current	2.1 A
Rated rotor voltage	2 V
Rated rotor current	0.26 A
Rotor speed	1200 rpm
p	6 poles
R_{gs}	0.414 Ω
L_{gs}	1.982 mH
R_{gr}	0.102 Ω
L_{gr}	18.5 mH
R_1	0.294 Ω
R_2	0.156 Ω
L_1	1.390 mH
L_2	0.74 mH
L_m	48 mH
N_s/N_r	0.02
Mechanical load	3 Nm
Moment of inertia	0.5 kg·m²

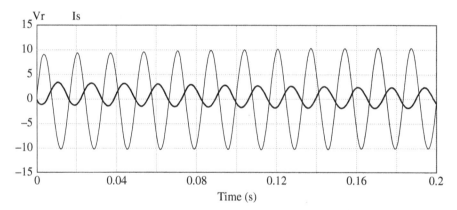

FIGURE 7.10   Output voltage of a DFIM connected to distribution network.

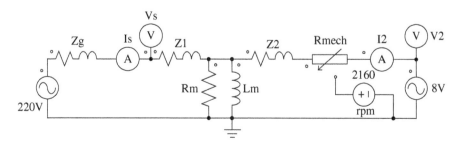

FIGURE 7.11   Per-phase model of a DFIM connected to the distribution network.

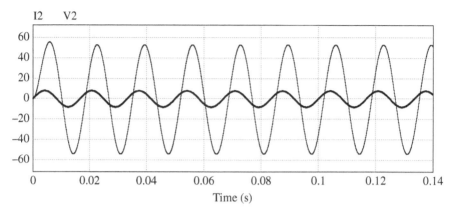

FIGURE 7.12   Output phase voltage of a DFIM connected to a distribution network.

TABLE 7.5   Parameters of the DFIM

Parameter	Magnitude
Rated stator voltage	220 V
Rated stator current	25.1 A
Rated rotor voltage	8 V
Rated rotor current	28.3 A
Rotor speed	2160 rpm
p	4 poles
R_g	2.141 Ω
L_g	6.85 mH
R_1	0.622 Ω
R_2	0.432 Ω
L_1	0.023 mH
L_2	0.032 mH
L_m	48 mH

The parameters used in Figure 7.11 are listed in Table 7.5.

7.6   DC MOTOR POWERING THE SHAFT OF A SELF-EXCITED INDUCTION GENERATOR

Some types of rotating loads demand for a good speed control. In laboratory tests, it is very common to use a DC motor driving a loaded SEIG to keep constant its load frequency. A good simulation like the example illustrated by the circuit shown in Figure 7.13 can help in establishing and sort practical restrictions that can be imposed on the DC machine. Notice the small capacitance of 1 nF used in parallel across the excitation bank of capacitors to establish the initial residual magnetism in the IM core without which the simulated SEIG does not start. The theory behind self-excitation of induction generators is available in Refs. [1, 6, 7].

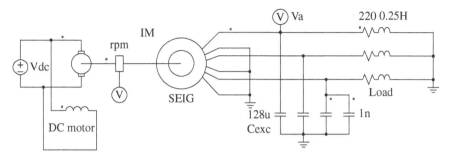

FIGURE 7.13   DC motor driving a loaded SEIG.

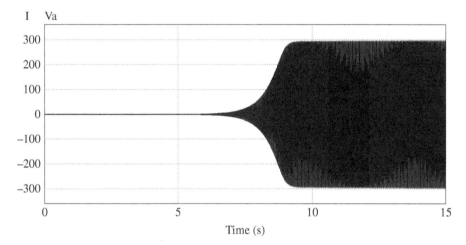

Time (s)

FIGURE 7.14   SEIG output voltage driven by a DC motor.

TABLE 7.6   Parameters of the SEIG

Parameter	Magnitude
Rated stator voltage	220 V
Rated stator current	15 A
Rated rotor speed	1800 rpm
Moment of inertia	0.4 kg·m²
p	4 poles
R_1	0.294 Ω
R_2	0.296 Ω
L_1	13.90 mH
L_2	13.90 mH

The simulation of a DC motor driving a SEIG establishes an output voltage as plotted in Figure 7.14 with parameters of the IM listed in Table 7.6 and the DC motor parameters listed in Table 7.7. This machine takes about 9 s to be fully excited.

TABLE 7.7   Parameters of the DC Motor

Parameter	Magnitude
Rated terminal voltage	120 V
Rated field current	1.6 A
Armature current	10 A
Armature resistance (R_a)	0.5 Ω
Armature inductance (L_a)	0.01 mH
Field resistance (R_f)	0.5 Ω
Field inductance (L_f)	0.01 V
Rated rotor speed	1800 rpm
Moment of inertia	0.4 kg·m^2

7.7   MODELING A PERMANENT MAGNET SYNCHRONOUS MACHINE (PMSM)

A full-fledged wind-turbine permanent-magnet synchronous generator connected to the grid with integrated load control is depicted in Figure 7.15. The theory behind such a complete wind energy system is available in Refs. [1–3]. A complete description of all the modeling and control aspects for such a system is too extensive for this book, but the reader is encouraged to study these (Refs. [4, 5]) in order to understand a more comprehensive picture of an overall simulation study for a PMSG. The permanent-magnet machine can be simulated based on the PSIM block, indicated in Figure 7.16. Table 7.8 shows the parameters used for the simulation model of Figure 7.14, and Table 7.9 lists the parameters of the shaft load. Figure 7.17 shows the voltage and angular speed of a PMSM connected to a mechanical load.

7.8   MODELING A SATURATED TRANSFORMER

Simulating a transformer with a PSIM block is very simple, as indicated by Figure 7.18, where it shows the setup of a saturated transformer feeding an RL load. The parameters of the transformer and load are listed in Tables 7.10 and 7.11. Figure 7.19 is the start-up plotting of the output phase voltage and current through phase "a."

7.9   LABORATORY PROJECT: TRANSIENT RESPONSE OF A SINGLE-PHASE NONIDEAL TRANSFORMER FOR THREE TYPES OF POWER SUPPLY—SINUSOIDAL, SQUARE WAVE, AND SPWM

A dynamic model for a single-phase transformer can be developed using the internal magnetic fluxes and can be valid for transient analysis. In this section, it is used the state-space flux-based equations of a single-phase transformer presented in the assignment description to develop its dynamic model based on its equivalent circuit for the transformer depicted in Figure 7.20:

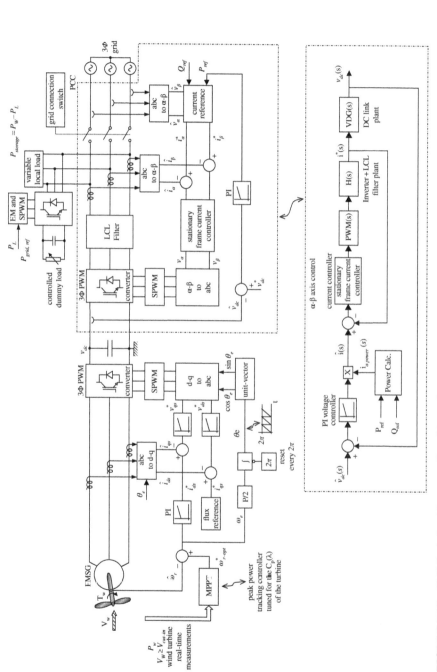

FIGURE 7.15 PMSG wind turbine controller with maximum power optimization and a back-to-back double-PWM grid-connected inverter with storage/load management.

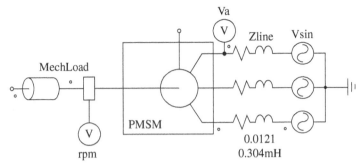

FIGURE 7.16 PMSM connected to a mechanical load.

TABLE 7.8 Parameters of the PMSG and AC Source

Parameter	Magnitude
Rated voltage	220 V
Rated current	8 A
Rotor speed	100 rpm
p	12 poles
R_{line}	0.012 Ω
L_{line}	0.304 mH
R_{stator}	0.530 Ω
L_d	22.6 mH
L_q	67.0 mH
V_{pk}/krpm	1.87 V/rpm
Inertia torque (J)	0.128 kg·m²
Shaft time constant	0.1 s

TABLE 7.9 Parameters of the Mechanical Load

Parameter	Magnitude
T_c	Nm
K_1	0.133 Nm/rad
K_2	0.012 Nm/rad²
K_3	0.001 Nm/rad³
Moment of inertia (J)	0.014 kg·m²

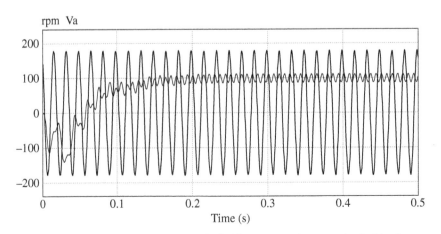

FIGURE 7.17 Voltage and rpm of a PMSM connected to a mechanical load.

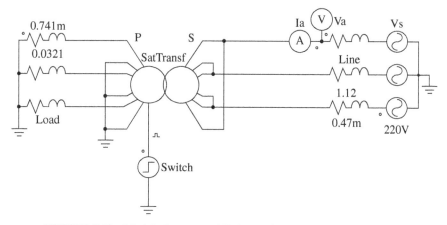

FIGURE 7.18   Model of a saturated Y–Δ transformer connected to the grid.

TABLE 7.10   Parameters of Saturated Transformer, Line and Load

Parameter	Magnitude
Rated voltage	150 V
Operating frequency	60 Hz
R_p	0.0268 Ω
L_p	0.04922 mH
R_s	0.0186 Ω
L_s	0.04922 mH
R_m	626.7 kΩ
I_m	67.0 mH
Phase A residual flux	0.6
Phase B residual flux	−0.3
Phase C residual flux	−0.3
N_p	1
N_s	0.1879
Per phase load resistance (R_L)	32.1 μΩ
Per phase load inductance (L_L)	104 μH
Per phase line resistance (R_{line})	1.12 mΩ
Per phase line inductance (L_{line})	4.7 uH

TABLE 7.11   Saturation Curve

I_m	L_m
0.75	386
984	0.59
1968	0.59

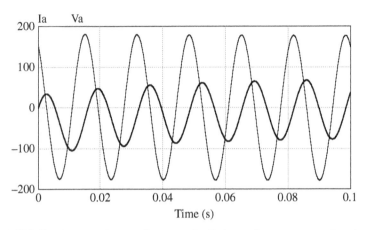

FIGURE 7.19 Output voltage and current of a Y–Δ transformer connected to the grid.

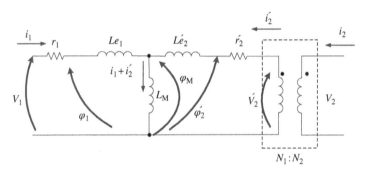

FIGURE 7.20 Equivalent circuit of a single-phase transformer indicating the magnetic fluxes.

$$V_1 = r_1.i_1 + \frac{d}{dt}\varphi_1 \tag{7.5}$$

$$V_2' = r_2'.i_2' + \frac{d}{dt}\varphi_2' \tag{7.6}$$

$$\varphi_1 = Le_1.i_1 + \varphi_M \rightarrow i_1 = \frac{\varphi_1 - \varphi_M}{Le_1} \tag{7.7}$$

$$\varphi_2' = Le_2'.i_2' + \varphi_M \rightarrow i_2' = \frac{\varphi_2' - \varphi_M}{Le_2'} \tag{7.8}$$

Substituting i_1, and i_2' on (7.5) and (7.6),

$$\frac{d}{dt}\varphi_1 = V_1 - r_1\left(\frac{\varphi_1 - \varphi_M}{Le_1}\right) \tag{7.9}$$

$$\frac{d}{dt}\varphi_2' = V_2' - r_2' \left(\frac{\varphi_2' - \varphi_M}{Le_2'} \right) \tag{7.10}$$

we need to calculate φ_M in order to replace it in (7.7) and (7.8):

$$\varphi_M = L_M \left(i_1 + i_2' \right) = L_M \left[\frac{\varphi_1 - \varphi_M}{Le_1} + \frac{\varphi_2' - \varphi_M}{Le_2'} \right] = \varphi_1 \frac{L_M}{Le_1} - \varphi_M \frac{L_M}{Le_1} + \varphi_2' \frac{L_M}{Le_2'} - \varphi_M \frac{L_M}{Le_2'} \tag{7.11}$$

$$\varphi_M \left(1 + \frac{L_M}{Le_1} + \frac{L_M}{Le_2'} \right) = \varphi_1 \frac{L_M}{Le_1} + \varphi_2' \frac{L_M}{Le_2'} \tag{7.12}$$

$$\varphi_M = \frac{Le_1 . Le_2'}{Le_1 . Le_2' + L_M Le_2' + L_M Le_1} \left[\varphi_1 \frac{L_M}{Le_1} + \varphi_2' \frac{L_M}{Le_2'} \right] \tag{7.13}$$

$$\varphi_M = \frac{L_M}{Le_1 . Le_2' + L_M (L'e_2 + Le_1)} \left[\varphi_1 L'e_2 + \varphi_2' Le_1 \right] \tag{7.14}$$

$$\frac{V_2'}{V_2} = \frac{N_1}{N_2} = \frac{i_2}{i_2'} \tag{7.15}$$

$$i_2 = -\frac{N_1}{N_2} i_2' \tag{7.16}$$

$$V_2 = \frac{N_2}{N_1} V_2' \tag{7.17}$$

If a resistive load is connected to the secondary, the output voltage becomes $V_2 = -R_L \cdot i_2$. The model is depicted on Figure 7.21.

In the webpage dedicated to the simulation examples of this book, there is a MATLAB/Simulink-based model available where the reader can study the connection to a 240 V, 60 Hz distribution utility grid. Assume that this transformer has the following parameters: $r_1 = 0.22\,\Omega$, $r_2' = 0.13\,\Omega$, $L_{1_1} = 148.5\,\mu H$, $L_{1_2} = 142.5\,\mu H$, $L_m = 1.655\,\mu H$. It is used to step down the input voltage to a secondary voltage of 120 V, connected to a resistive load $R_L = 7.5\,\Omega$. The diagram of Figure 7.21 and the Simulink block diagram in Figure 7.22 show how to simulate a transient model for a transformer using equations for simulating the transformer differential equations 7.5–7.17. It is possible to plot input primary current, magnetizing current, and output secondary current and evaluate the efficiency; see results of Figure 7.23

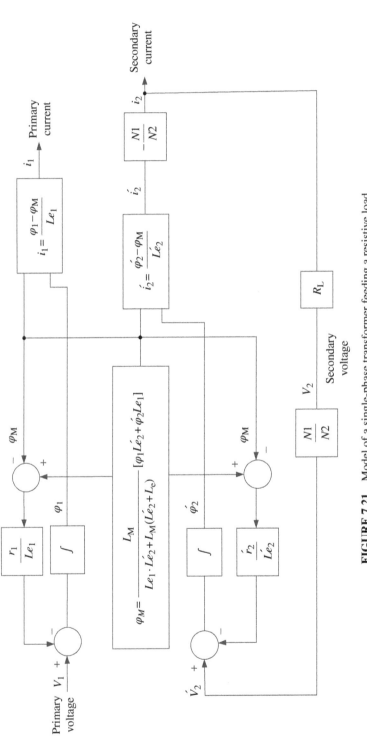

FIGURE 7.21 Model of a single-phase transformer feeding a resistive load.

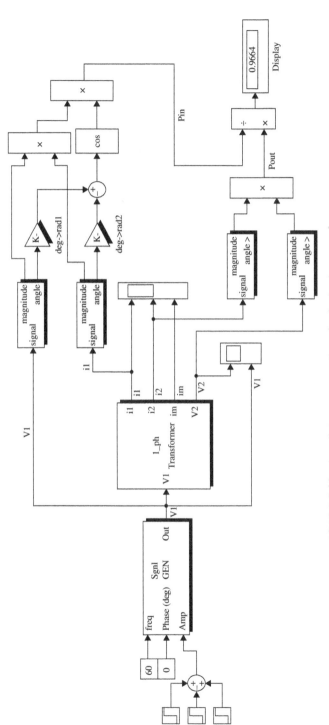

FIGURE 7.22 Simulink-based modeling of a single-phase transformer.

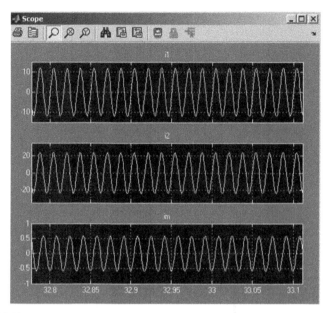

FIGURE 7.23 Input primary current, output secondary current, and magnetizing current.

related to Simulink model in Figure 7.22. Based on those results, the efficiency is evaluated to be 97%.

It is also possible to study the transient response for a voltage dip of 10% for five cycles, simulating as long as it is needed to capture the transient and study the transient response for a voltage surge of 10% for five cycles, simulating as long as it is needed to capture the transient. Based on simulation results, it is observed that the occurrence of voltage sags and swells at the primary side directly affects the input and output currents as well as output voltage. Also, it can be seen that there is a huge DC current on the magnetizing current, and its time constant is very large. One reason for that is the large value of magnetizing reactance, because, as we know, the time constant of a RL circuit is directly proportional to L. So, the magnetizing current cannot follow the system transients because it is so sluggish, with a large time constant and settling time. On the other hand, as an advantage of having a large magnetizing reactance, a high efficiency of the system (i.e., 0.97) can be mentioned.

With this modeling approach, it is possible to simulate a transformer in conditions that evaluate its magnetizing input current and estimate its power factor (Figures 7.24 and 7.25).

In this part, we would like to impose some specific conditions on the system in order to evaluate the magnetizing current. To study the magnetizing current, we need to operate it at no-load condition. We can reach this condition by increasing the load resistance, because in that condition, $i_2 = 0$ and as a result $I'_2 = 0$; hence, $i_1 = i_m$ as observed in Figure 7.26. In this section it is possible to apply these conditions and study the system performance to verify the relationship for the primary, secondary, and magnetizing currents.

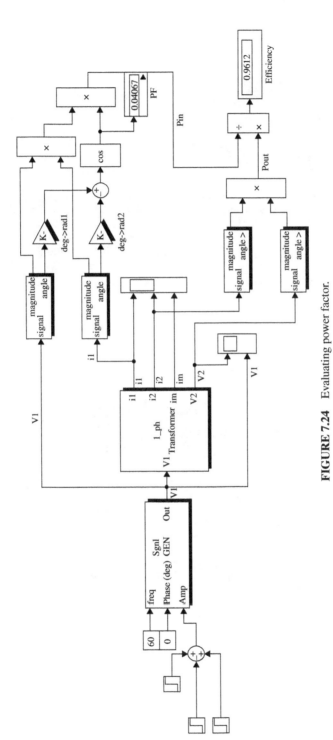

FIGURE 7.24 Evaluating power factor.

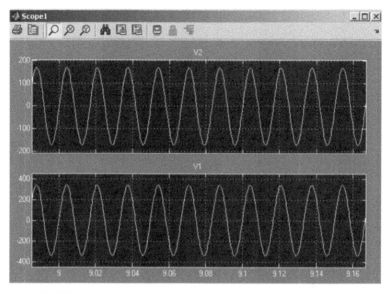

FIGURE 7.25 Primary and secondary voltages (exactly in phase).

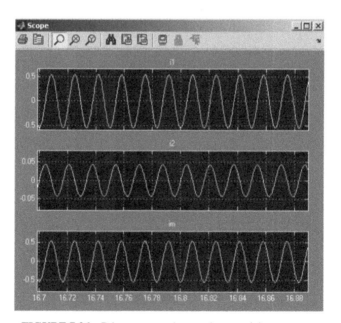

FIGURE 7.26 Primary, secondary, and magnetizing currents.

Based on the model, the PF of the magnetizing branch is equal to 0.0407 at steady-state condition. In this case, $R_{load} = 4000\,\Omega$. The efficiency (0.96) and power factor (0.407) of the system have interesting values at no-load condition (i.e., this part). They show that the efficiency at a no-load condition is lower in respect to full-load condition but it is still high enough to make it acceptable. The power factor is very close to zero. That means the circuit is highly inductive or the resistors are negligible in respect to reactors. Figure 7.27 shows how this transformer can be connected to single-phase inverter, with a DC-link voltage of 600 V. Simulate a square-wave voltage (60 Hz) across the primary side, with similar transformer parameters and load as previously mentioned in this chapter, and compare a sinusoidal excitation with a square-wave excitation performance. Evaluate its response and performance for this square-wave operation in Figures 7.28 and 7.29.

The simulation results show that the output waveforms have very high harmonic contents. As a result, this system does not have a proper harmonic performance and may cause serious damages to transformer (i.e., heating, losses), and proper transformer design and assembling should be done for real-life application.

There is an important point regarding the efficiency and power factor shown in the aforementioned Simulink model. In calculating these measures, we took the fundamental component of voltages and currents and did not consider harmonics. We know that the appearance of harmonics in a system can affect these measurements and will decrease them. So, the real efficiency and power factor are much lower than that shown previously. The waveform of the magnetizing current seems like a sawtooth waveform which is derived by integrating a pulse waveform. In conclusion, this system is not recommended.

Figure 7.30 shows a single phase dc-link transistor converter for a transient modeling of a transformer. Implement a simulation of the inrush current and study a way to soften it to 10% of its maximum natural value.

Using previous simulation studies, implement a three-phase sinusoidal PWM (SPWM), operating at $f_{PWM} = 5\,kHz$, in order to control a three-phase load connected through a transformer with the parameters listed prevously. Command your three-phase transformer with a modulation index of 20, 40, 80, and 100%. Critically analyze the operation and compare the transformer response for the (i) sinusoidal operation, (ii) square-wave operation, and (iii) SPWM operation. Observe Figures 7.31, 7.32, 7.33, 7.34, and 7.35.

This simulation-based study allows the understanding of operating principles of a balanced and symmetrical three-phase transformer with SPWM-derived variable input voltage. By changing the modulating index, the generated voltage pulses vary and can be programmed, for example, the envelope or amplitude is kept fixed. Therefore, the output voltage of the transformer can be controlled. Similar studies can be implemented for an induction or a synchronous machine (Figure 7.35).

7.10 SUGGESTED PROBLEMS

1. Demonstrate with simulation studies that the equivalent of an IM connected to the grid (Figure 7.1) can be operated either as a motor or a generator. Can you estimate the overall losses in both cases using PSIM measurements?

2. Develop a control system in PSIM to guarantee a 60 Hz frequency across the load in the asynchronous power plant illustrated in Figure 7.10. What would happen if

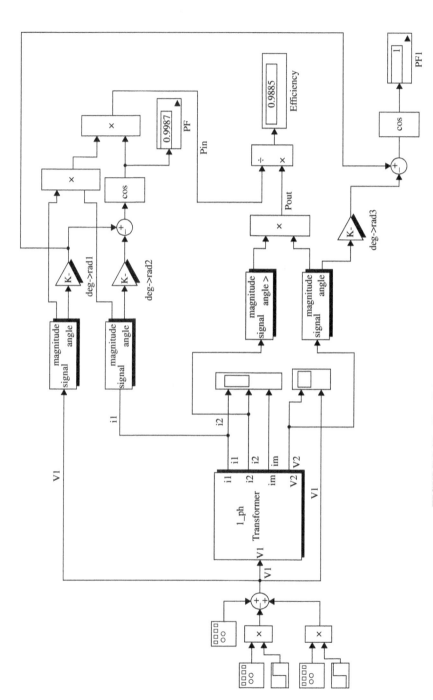

FIGURE 7.27 Simulink model for square-wave excitation.

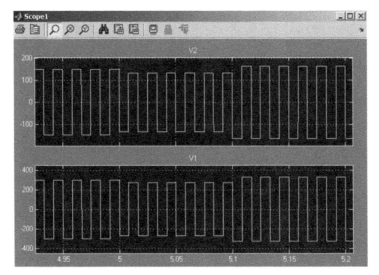

FIGURE 7.28 Primary and secondary voltages for square-wave supply.

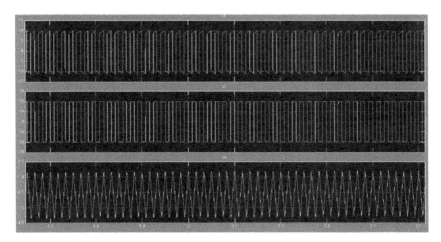

FIGURE 7.29 Primary, secondary, and magnetizing currents.

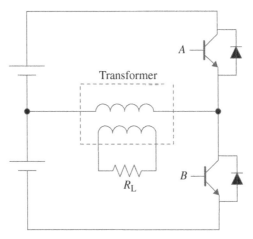

FIGURE 7.30 Single-phase split DC-link transistor converter for a transient modeling of a transformer.

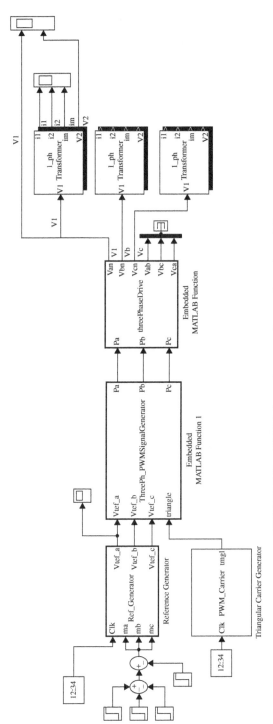

FIGURE 7.31 SPWM three-phase transformer dynamic modeling.

FIGURE 7.32 Primary and secondary voltages on a long time span (15 s).

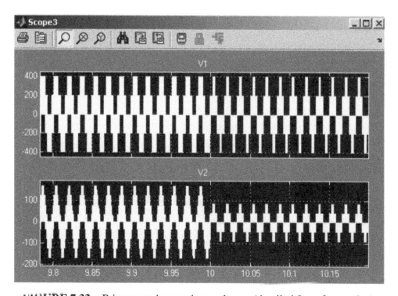

FIGURE 7.33 Primary and secondary voltages (detailed for a few cycles).

the load increases by 10%? Can you develop a variable excitation bank of capacitors to assure a reasonably constant output voltage?

3. Modify the circuit presented in Figure 7.1 to include measurement of the power injected into the AC grid, efficiency, and losses.

4. Explain the advantages a DFIG can offer related to increase torque, harmonic reduction, and speed control for maximum torque. Develop a simulation study for a DFIGURE.

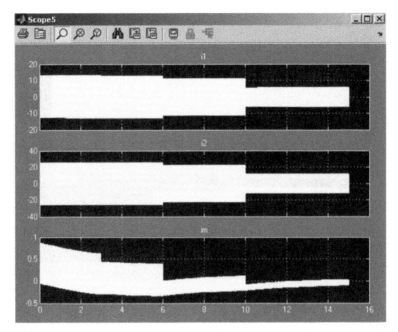

FIGURE 7.34   Primary, secondary, and magnetizing currents on a long time span (15 s).

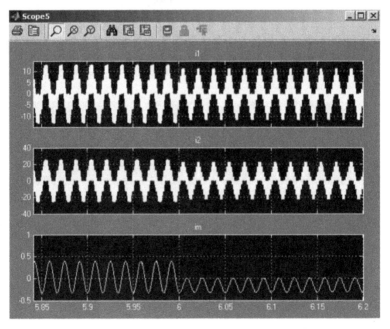

FIGURE 7.35   Primary, secondary, and magnetizing currents (detailed for a few cycles).

5. Develop a hill-climbing control (HCC) to tune the generator showed in Figure 7.5. Develop a simulation study to observe (i) maximum power, (ii) maximum efficiency, (iii) maximum voltage, (iv) maximum current, and (v) minimum losses. Develop a program that is capable of consider a weighted optimization index of all five points mentioned previously for an averaged optimization of all the factors.

6. Use the self-excited induction generator shown in Figure 7.12 to develop a circuit for obtaining the electrical characteristics of induction machines working as generators.

7. Simulate a permanent magnet (PM) synchronous generator-based wind energy system as portrayed on Figure 7.15.

REFERENCES

[1] SIMÕES, M.G., FARRET, F.A. and BLAABJERG, F., Small wind energy systems, electric power components and systems, Electric Power Components & Systems, vol. 43, pp. 1388–1405, 10.1080/15325008.2015.1029057, 2015.

[2] FARRET, F.A. and SIMÕES, M.G., Integration of Alternative Sources of Energy, IEEE-Wiley Interscience/John Wiley & Sons, Inc., Hoboken, 2006.

[3] SIMÕES, M.G. and FARRET, F.A., Modeling and Analysis with Induction Generators, 3rd edition, CRC Press, Boca Raton, 2015.

[4] BU-BSHAIT, A.S., SIMÕES, M.G., MORTEZAEI, A. and BUSARELLO, T.D.C., "Power quality achievement using grid-connected converter of wind turbine system", IEEE Industry Applications Society Annual Meeting (IAS), Dallas, IEEE, October 18–22, 2015.

[5] HARIRCHI, F., SIMÕES, M.G., BABAKMEHR, M., AL-DURRA, A. and MUYEEN, S.M., "Designing smart inverter with unified controller and smooth transition between grid-connected and islanding modes for microgrid application", IEEE Industry Applications Society Annual Meeting (IAS), Dallas, IEEE, October 18–22, 2015.

[6] CHAPMAN, S.J., Electric Machinery Fundamentals, WCB McGraw-Hill, Boston, ISBN-13: 978-0073529547, ISBN-10: 0073529540, 2011.

[7] FITZGERALD, A.E., KINGSLEY, C., JR. and UMANS, S.D., Electric Machinery, 7th edition, McGraw-Hill Education, New York, ISBN-10: 0073380466; ISBN-13: 978-0073380469, 2014.

FURTHER READING

DEL TORO, V., Basic Electric Machines, Prentice Hall, ISBN-10: 0130601462, ISBN-13: 978-0130601469, 1989.

FEDÁK, V., BALOGH, T. and ZÁSKALICKÝ, P., Dynamic Simulation of Electrical Machines and Drive Systems Using MATLAB GUI, Ministry of Education of Slovak Republic Under the Contract KEGA 042TUKE-4/2012 "Teaching Innovation in Control of Mechatronic Systems", Slovakia, pp. 317–342, 2012.

LOGUE, D. and KREIN, P.T., "Simulation of electric machinery and power electronics interfacing using MATLAB/Simulink", The 7th Workshop on Computers in Power Electronics, 2000. COMPEL 2000, Blacksburg, IEEE, doi:10.1109/CIPE.2000.904688, pp. 34–39, July 16–18, 2000.

8

STAND-ALONE AND GRID-CONNECTED INVERTERS

8.1 INTRODUCTION

Smart distribution systems require that existing and new assets, such as distributed generation (DG) and distributed energy storage (DES) units, become adaptable for provision of supplying not only electricity but also ancillary services for the local grid. Those DG and DES units require power electronic front-ends capable of functioning in multiple quadrants and capable of delivering electrical power under good quality indices. Some ancillary services could be made available with power electronics to support the local grid. Therefore, a control system will have to handle the instantaneous power balance as well as the long-term energy requirements.

A control system when properly designed for a power electronic energy conversion system will observe the outputs and the internal states and make sure that the overall system will follow a desired set point. For example, we may design general control systems for (i) automatic control, for example, control of room temperature; (ii) remote control, for example, satellite orbit correction; and (iii) power amplification, for example, control system to impress circuits with sufficient electrical power to turn on heavy machines. As regarding control systems for stand-alone and grid-connected inverters, the following control attributes and features are the most usual ones:

- Voltage regulation
- Frequency regulation

Modeling Power Electronics and Interfacing Energy Conversion Systems, First Edition.
M. Godoy Simões and Felix A. Farret.
© 2017 John Wiley & Sons, Inc. Published 2017 by John Wiley & Sons, Inc.

- Provision of spinning reserves and standby service
- Power quality enhancement
- Power factor correction
- Instantaneous power compensation during load transients
- Intentional islanding for demand response

A power electronic inverter operating as a front end for a DG resource, such as a photovoltaic (PV), or wind, or fuel cell, or hydropower system, can have multifunction features such as those previously mentioned and can be designed to have a high level of intelligence. As part of a smart grid system, smart metering and sensor systems are expected to provide large amounts of information for management and control purposes—demand side management (DSM) and demand response (DR) can be implemented and supported by two-way smart meters and smart sensors on equipment, opening further possibilities for real-time and scalable market mechanisms. Near real-time information allows utilities to manage the power network as an integrated system, actively sensing and responding to changes in power demand, supply, costs, and quality across various locations.

Grid-connected inverter interfacing methods have historically evolved from frontend pulse-width-modulation (PWM) rectifiers used in regenerative machine drives, where instead of dissipating power on dynamic break (from deceleration or torque reversal transient operation) the efficiency is improved by pumping power back to the grid. A grid-connected inverter has same control principles as a PWM rectifier. However, further flexibilities are required when such systems are injecting power into the grid from renewable energy sources (PV, hydro, or wind) or from fuel cells because an overall balance of system must be provided.

When the inverter is in stand-alone mode, a voltage control will maintain the output voltage with a clean sinusoidal output waveform. When the inverter is connected to the grid, the inverter must have a current loop control in order to inject a sinusoidal current exactly 180° phase shift with the grid voltage (for unity power factor operation), or the current phase shift can be programmable for leading or lagging power factor operation. For this reason, synchronization with the grid must be provided using a phase-locked loop (PLL) with inner current control, sometimes associated with an outer power (active/reactive) control. If the inverter has automatic transition from grid-connected to islanded or vice versa, the system must be capable of providing such switch in the current/voltage or voltage/current closed-loop control. There is also a need for detection of utility grid blackout (to disconnect the system from the grid and avoid feeding a faulty line) and then a return back without major transients.

For the grid-connected interfacing, typically it is used a voltage source inverter (VSI), such as one shown in Figure 8.1. It is always necessary to have an impedance between the inverter and the grid, such as a decoupling inductor, or interphase reactors, or another type of filter. Voltage source converters connected to the grid have been demonstrated to have improved performance with LCL filters. In addition, the control system for injecting power into the grid can be designed (by analogy) to

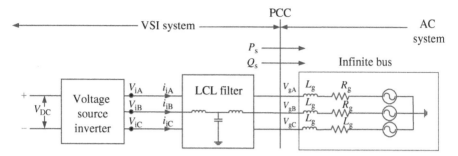

FIGURE 8.1   Typical voltage source inverter (VSI) connected to the utility grid.

resemble machine control techniques using cascaded loop control, where the voltage is commanded in the outer loop (similar to a machine speed in a drive system) and current in the inner loop (similar to a machine torque in a drive system). Back-EMF cross-coupling motional voltages must be compensated—a grid-connected inverter presents a cross-coupling voltage drop (a system reactance multiplied by current term from one d–q axis disturbing the other d–q axis). Therefore, a proportional–integral (PI) controller must be properly designed to optimize such cross-coupling effects. Another possibility is by using P+resonant controllers to improve dynamic performance of the current control loop in the stationary dq frame or in the ABC frame. The advantage of P+resonant is that PLL synchronization is not necessary, because the control is made on the d–q stationary reference frame, instead of the d–q rotating reference frame.

Three-phase systems are easily controlled with decoupled d–q or p–q instantaneous power theory. Those strategies have been developed for long time already for three-phase machine control. However, single-phase grid-connected inverters (which are the most common in power distribution systems) do not have a proper mathematical formulation. For single-phase inverters, typically a hysteresis-band (bang–bang) control, or an approximated solution that considers a phase shifted by 90° on the main voltage, or Hilbert transformation can be used. Those quasi-transient decoupled d–q or p–q methodologies are applied on artificial formulation [1, 2].

Inverters for DG systems are designed to be either utility interactive or utility independent. Utility interactive inverters are controlled to behave as a current source, delivering power to the utility with near-unity power factor, with multi-functionalities. Utility-independent inverters behave as voltage sources, where the load on the inverter prescribes the phase difference between the output voltage and the load current.

Figure 8.1 shows a boundary defined as point of common coupling (PCC). The term "point of common coupling" became widely used after the release of IEEE 519, "Recommended Practices and Requirements for Harmonic Control in Electrical Power Systems," which defined it as "the interface between sources and loads on an electrical system." Most power quality professionals originally thought the PCC to be at the service transformer. This definition/location benefits the electric utility, which does not want harmonic currents above a certain level from a connected

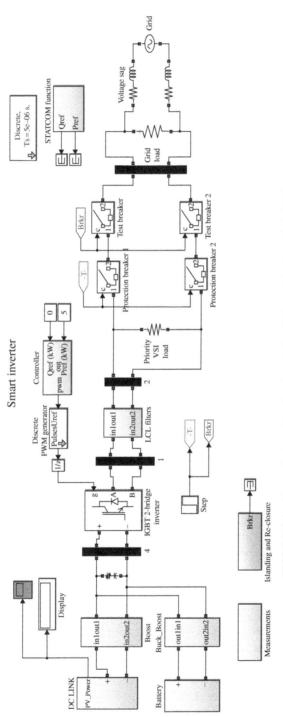

FIGURE 8.2 Power Systems Toolbox-based Simulink model of a grid-tied PV system.

customer. But there are other possibilities, for example, if the service equipment has a backup engine gen-set or similar power source, the interface to this power source becomes the PCC when the transfer switch connects to that source and when the normal electrical supply disconnects. In addition, if an electrical system equipment can be subdivided into smaller sections each served by a transformer installed as a separately derived AC system, you would have a subset of PCCs across each transformer.

Feeders also have an effect on the type of power quality problem referred to earlier because they have impedance that increases as a function of their length. Therefore, a panel board that serves nonlinear loads will exhibit greater voltage waveform distortion at its buses, as compared with the upstream end of the feeder where the PCC exists. The PCC is important because it is the point where the interaction of the served nonlinear load's distorted current waveform may affect the voltage waveform distortion. This voltage is then provided to the bus or feeder served by the power source, propagating to all served downstream loads as a power quality problem related to the distorted voltage waveform.

A stand-alone inverter system simulation is depicted in Figure 8.2 implemented with the Power Systems Toolbox and Simulink. There is a *voltage controller*, where the reference voltage is *Vd_ref(pu)*, which is compared to the peak voltage in per-unit *Vi_abc(pu)*. The voltage controller has a sinusoidal reference generator *Sin_Cos* for the *dq0–abc conversion* for generating the output reference *Vref_abc* that will serve as input for the *discrete PWM generator*. The following sections have examples for implementing interfacing control methods for grid-connected inverters, and a laboratory project example is discussed. There are several other interfacing methods reported in the literature, and they are left for the interested reader for further evaluation [3–8].

8.2 CONSTANT CURRENT CONTROL

In the grid-connected mode, the inverter is usually designed to supply constant current output in order to connect to the voltage source grid utility. The constant current control is usually implemented with reference frame transformation from three phase to stationary frame (*abc/dq*) and with the unity vector synthesized by a PLL. A reference frame (Figure 8.3) rotates synchronously with the grid. Equation 8.1 describes the grid dynamics, which is used for the feedback control. Typically PI controllers whose output generates voltage on the reference frame are used to impress the variation of voltage on each axis, that is, Δe_d and Δe_q. Figure 8.4 shows a control diagram where the output voltages are transformed to set up references for a PWM controller, which by its turn commands the gating pulses of the transistor-based inverter. A PLL is required to synthesize the unit vector, that is, sine and cosine waves that allow synchronization with the utility grid:

$$L\frac{d}{dt}\begin{bmatrix} i_d \\ i_q \end{bmatrix} = \begin{bmatrix} R & -\omega L \\ \omega L & R \end{bmatrix}\begin{bmatrix} i_d \\ i_q \end{bmatrix} + \begin{bmatrix} \Delta e_d \\ \Delta e_q \end{bmatrix} \tag{8.1}$$

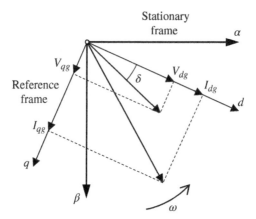

FIGURE 8.3   Stationary and reference frame for grid variables.

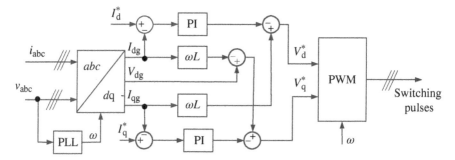

FIGURE 8.4   Current-controlled PWM inverter with d–q decoupled reference frame with PLL grid synchronization.

In Equation 8.1 variables i_d, i_q are the d- and q-axis current components, and Δe_d and Δe_q are the voltage instantaneous differences between the PCC and the inverter output voltage of d and q axis components.

8.3   CONSTANT P–Q CONTROL

The generalized theory of the instantaneous reactive power (p–q theory) in three-phase circuits has been proposed by Akagi in 1983 [9], and it was used for the first time to control grid-connected inverters in 1991 by Ohnishi [10]. Equations 8.2 and 8.3 indicate calculation of variables p and q for ideal conditions. There are some advantages of instantaneous power-based controllers: (i) power calculation does not vary by any transformation; (ii) it is directly related to energy-constrained sources, such as PV and wind; and (iii) it allows instantaneous power balance with the utility grid. Figure 8.5 shows an implementation diagram where active power builds up the q-reference value for current, which is compared to the instantaneous grid q-current,

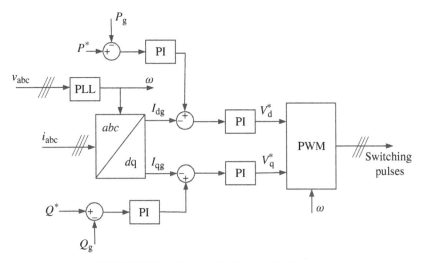

FIGURE 8.5   Constant P–Q control technique.

whereas the reactive power loop builds up the reactive power d-reference current value, which is compared to the instantaneous grid d-current. Therefore, a constant p–q control is an outer loop over a current loop control.

$$p = \frac{3}{2}\left[v_{ds}i_{ds} + v_{qs}i_{qs}\right] \tag{8.2}$$

$$q = \frac{3}{2}\left[v_{qs}i_{ds} - v_{ds}i_{qs}\right] \tag{8.3}$$

The main drawback of the p–q technique based on Akagi's instantaneous power theory is that the quantity "q" of Equation 8.3 is only considered for the reactive power for symmetrical, balanced, and nondistorted three-phase systems, that is, it is not valid for distorted three-phase systems. Typically, at least the third harmonic distortion is very common in electrical power distribution. For any other situation having zero sequence or distortions, this quantity must be called *imaginary power* (instead of reactive). In addition, the Akagi's formulation is valid only for three-phase systems. The current literature uses the Akagi's theory for single-phase systems with some kind of approximation. Therefore, using Akagi's theory does not properly address single-phase systems, neither instantaneous reactive power management. However, when a controller impress to maintain always $q=0$, then it will impose a good unity power factor operation for balanced three-phase systems.

8.4   CONSTANT P–V CONTROL

The grid-connected inverters make possible the control of voltage at the PCC. This technique is a slight variation of the previous one, where the voltage feedback is used to adjust the amount of reactive power (or d-axis voltage) required for the system, as

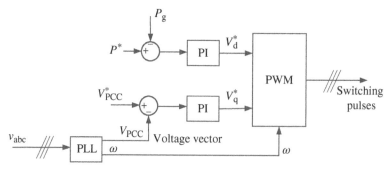

FIGURE 8.6   Constant P–V control technique.

indicated by Figure 8.6. Although it seems simple, there are some challenges in this approach, where the instantaneous peak voltage of the utility grid must be followed at the PCC having an impedance, usually inductive, between the PCC and the inverter that will affect the stability of the feedback control loop. The P–V control approach is the most suitable for voltage support utility services, but it is very difficult to define the stability range of operation by mathematical analysis, and very extensive real-time simulation and experimentation must be conducted in order to implement the hierarchical level of active power and voltage grid control.

8.5   IEEE 1547 AND ASSOCIATED CONTROLS

Control algorithms performing connection (and islanding) of a power electronic (PE)-based DG with the grid should be compliant with the IEEE 1547 family of standards [11]. Control functionalities must take into account a variety of input parameters such as frequency, phase, and d–q frame voltage of the grid and the PE side. The control algorithm must compare these inputs with the IEEE 1547 recommendations and generate appropriate signals with recommended critical timing to island or to reconnect the PE-based DG to the grid. A smart inverter must take into account the provision of at least one ancillary service in addition to the primary service [11, 12]. Otherwise, it would be only a "regular inverter."

A VSI can be defined as a multifunctional inverter if it has combined functional abilities such as (i) to supply power to local loads, (ii) supply power to other utility loads up to the rated capacity of the inverter, (iii) provide voltage support at the PCC of the utility, (iv) store energy in a local lead acid battery bank or other types of storage, and (v) provide control options to the consumer based on near real-time electricity information obtained from the utility through advanced metering devices. Smart inverter functionalities are beyond the recommendations of the current national technical standard for interconnecting DG sources to the grid—the IEEE Std. 1547 [11]. In addition, the use of reactive power in providing voltage support at the PCC should not be implemented without a proper authorization of the local utility company. Traditionally, voltage support and voltage sag correction in distribution systems are performed using utility capacitor banks. But with the advent of inverters with smart functionalities, the ability to regulate voltage at the PCC can be brought to the customer, particularly if that is completely within their own microgrid. The authors have not yet probed all the safety issues stemming from performing such voltage controls. For such, real-time communication with the local utility must be implemented. If real-time spot pricing of electricity is available from the local utility, by using an advanced metering device, the inverter control algorithm can determine an optimal operating mode, and the designer could enable the inverter to (i) schedule local loads and (ii) determine to either store energy locally or sell electrical power back to the grid. A possible simulation of a smart inverter is depicted in Figure 8.7.

The smart functionalities of an inverter connecting an energy source to grid must be aimed at (i) provision of real and reactive power support to local loads, (ii) provision of real and reactive power to grid loads up to the rated capacity, (iii) option to control voltage at the PCC during voltage sags, and (iv) decision-making ability aided by information of real-time pricing. Based on such functionalities, the inverter

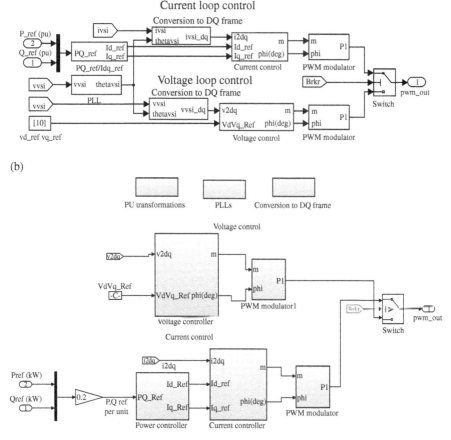

FIGURE 8.7 Multifunctional inverter. (a) Current control Simulink block diagram and (b) voltage control Simulink block diagram.

operation is governed by certain rules, which determine the mode of operation. To show a performance of the implemented circuit in Simulink at $t=0.5$ s, the inverter switches its status from grid-tied to stand-alone mode. Figure 8.8 shows that the voltage and current waveforms on the inverter side are out of phase, corresponding to the fact that the inverter is supplying active power with unity power factor. The

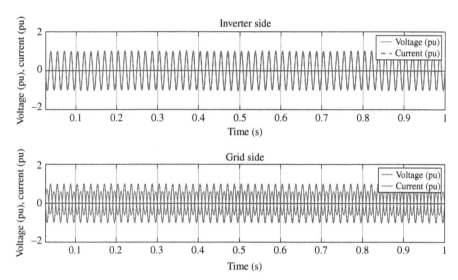

FIGURE 8.8 Active power injection, showing voltage and current on the grid side in grid-connected mode.

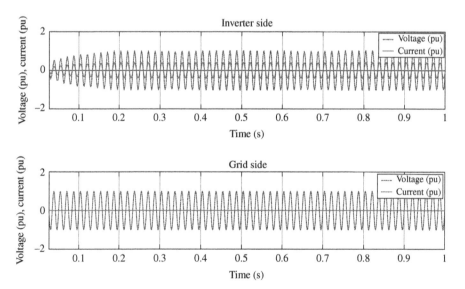

FIGURE 8.9 Stand-alone operation, showing voltage and current on the inverter side in islanded conditions.

voltage and current are both at 1 pu. Figure 8.9 shows the stand-alone operation, depicting voltage and current on the inverter side, in islanded conditions.

8.6 P+RESONANT STATIONARY FRAME CONTROL

A possibility of control implementation is by using the stationary reference frame [13] as shown in Figure 8.10. In this control structure, the grid currents are transformed to stationary reference frame using the $abc \rightarrow \alpha\beta$ transformation (also called Clarke's transformation). One very popular controller is the proportional-resonant (PR) controller for current regulation of grid-tied systems [14, 15] where the setpoint variables are sinusoidal. This controller achieves a very high gain around the resonant frequency (same as the grid frequency), imposing the sinusoidal output behavior and eliminating the steady-state error between the controlled signal and its reference [16]. Such kind of controller is derived from the *repetitive control* technique [17].

The PR controller does not work well for variable frequency operation such as for machine drive systems or when the utility system is a weak grid instead of a strong one. However, for typical small frequency variation of a typical modern power grid (<0.5 Hz), it can track really well. One advantage of this controller is that a PLL is not necessary, neither the trigonometric transformations from the stationary frame to the reference frame.

The controller with resonating terms can be implemented as indicated in Figure 8.11. In addition of working in three-phase inverter, this control method allows to calculate the current reference for single-phase inverter [18, 19].

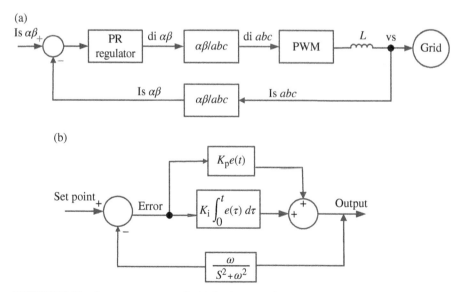

FIGURE 8.10 P+resonant approach. (a) Stationary reference frame control and (b) details of controller with resonating term.

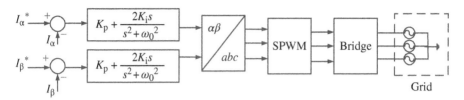

FIGURE 8.11   α–β stationary frame controller with P+resonant control in *s*-domain.

A typical approximated solution for the single-phase inverter control is to take the single-phase voltage as the α-component and have a 90° delay for that signal as the β-component. The α-component of the controller output should be taken as the output single-phase quantity to be fed back to the sinusoidal pulse width modulation (SPWM) unit. This method does not demand any transformation matrices, so it is easy to be implemented. Like its three-phase counterpart, it produces a rotating vector with the constant magnitude in rotational frame when the magnitude of the reference current is constant. However, it may not be optimized for transient conditions, because the 90° phase shift is only valid for steady-state conditions.

8.7   PHASE-LOCKED LOOP (PLL) FOR GRID SYNCHRONIZATION

A PLL system is commonly used for various signal applications such as radio and telecommunications, electrical motor control, and signal processing. In the last few years, PLL became important for power electronic applications, providing synchronization with the utility grid. PLL techniques can be adapted to work in a wide frequency range from a few hertz to orders of gigahertz. There are mainly three types of PLL systems for phase tracking: (i) zero crossing, (ii) stationary reference frame, and (iii) synchronous reference frame (SRF)-based PLL [20]. The SRF PLL is one with a good performance under distorted and non-ideal grid conditions, and it is applicable in both single-phase and three-phase systems [21].

A basic PLL configuration is depicted in Figure 8.12. The phase voltages v_{gA}, v_{gB}, and v_{gC} are obtained from sampled phase voltages. These stationary reference frame voltages are transformed to voltages V_d and V_q in a frame of reference synchronized to the utility angular frequency using α–β and *d–q* transformations.

The angle θ^* used in these transformations is calculated by integrating the frequency signal ω^* where the initial angle must be carefully reset as the initial condition in the integrator. If the frequency command ω^* is identical to the utility frequency, the voltages V_d and V_q appear as DC values depending on the angle θ^* [21].

The $\alpha-\beta$ or Clarke transformation allows to represent a three-phase system v_{gA}, v_{gB}, and v_{gC}, as a two-phase system $V\alpha$ and $V\beta$. The control in such $\alpha\beta$-frame has the feature of reducing the number of required control loops from three to two. However, the reference and feedback signals are sinusoidal functions of time in the $\alpha\beta$-frame. Therefore, in order to achieve a satisfactory performance and small steady-state

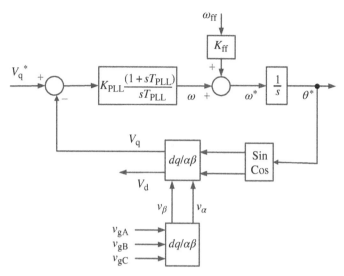

FIGURE 8.12   A PLL with dq–$\alpha\beta$ transformation.

errors in magnitude and phase, the compensator design is not trivial, and a P+resonant control can be used. The dq frame-based controller (using a Park transformation) keeps the signals as DC waveforms under steady-state conditions, and the transients are exponentials. Therefore, in the dq frame-based controller (with a Park transformation), it is possible to use compensators with simpler structures and lower dynamic orders, such as regular PI controllers.

A feedforward reference ($\omega_{ff} = 2\pi f_g$) is included to improve the initial dynamic performance in the PLL. The frequency f_g is the grid nominal frequency. Adding ω_{ff} helps to improve the start-up time of the PLL [21–23]. The utility grid is typically a stiff (strong) system with regard to their supply frequency, and there are very small variations depending on the droop control used by the utility for power flow, but such variation can be usually neglected for the distribution grid. If there is a deviation in the supply frequency, that will cause the phase angle error to increase. The PI regulator naturally works to bring such error to zero. The reaction to frequency fluctuations is completely predictable by the closed-loop response of the PLL system. The feedforward term ω_{ff} applied through the gain K_{ff} facilitates the function of the regulator. If the supply frequency is changing in the microgrid where the inverter is installed, for example, with stand-alone diesel power generators, there will be a tracking error in the phase angle θ (as long as the frequency is changing). If the change in frequency is predictable, the tracking error can be eliminated by the feedforward term. Some fine-tuning can be done initially in the simulation and adjusted in the real application. If the change in frequency is not predictable, an additional integral term may be used in the PI regulator, similar to an antiwindup PI scheme.

8.8   LABORATORY PROJECT: SIMULATION OF A GRID-CONNECTED/ STAND-ALONE INVERTER

The goal of this project is to design a controller for power electronic interfaces comprised of a DC/DC boost converter and a grid-connected inverter for a PV energy source. The input voltage of the DC/DC boost converter is 80 V, and the output voltage is 380 V. A type III controller is implemented for the power converter. Theoretical results are evaluated using simulation in Matlab and PSIM software. In order to find suitable values for the controller, it used a manual design to find the values of the components of this compensator. Moreover, a controller designed for grid-connected inverter uses Clarke and Park transformations (d–q theory) in order to define the control signals. The advantage of using this methodology is that a regular PI controller works well and it is easy to understand and possible to design using classic control methods.

A grid-connected inverter must be capable of identifying a grid fault or a blackout and change its control mode from a current controller to a voltage controller. Table 8.1 shows the parameters for a converter with a DC link voltage of 380 V supplying an output voltage of 120 V_{AC} (RMS). The inverter has a LCL filter connected to the grid, which must be properly designed [24, 25].

The control of a grid-connected inverter, particularly when used for microgrid applications, involves many challenging issues. In order to operate a microgrid properly in different operation modes and during operation mode transitions, good power management strategies including real and reactive power control, frequency and voltage regulation, synchronization, and load demand matching should be developed.

This laboratory project covers mostly the power flow control in the grid-connected operation mode. In this mode, the main function of a DG unit is to impress the real and reactive power at the PCC. The real power reference is given from a higher-level energy management controller. If the energy source is constrained, such as in a wind turbine or in a PV array, a coordination with the maximum power point tracking (MPPT) controller should be designed. The reactive power reference can be maintained at zero for unity power factor operation—typically it is the way that most commercial inverters operate, or it can be commanded according to the grid reactive power or voltage requirement (it requires either droop control or coordination with the local utility).

Although the inverter control method used in this project is based on the d–q theory (Clarke and Park transformations), there are other ways to implement a high

TABLE 8.1   Inverter Parameters

Input DC voltage (V_{DC})	380 V
Output AC voltage (V_{OUT})	120 V_{rms}
Inductor of LCL (L)	1 mH
Capacitor of LCL (C)	100 μF
Switching frequency (f_{SW})	15 kHz
Grid frequency (f_G)	60 Hz

performance control. Readers are encouraged to study further other possibilities. The three-phase AC voltages and currents to two DC variables can be transformed, and PI controllers can be used to impress the command PWM for the inverter. Therefore, the whole implementation can retrofit any DSP or microcontroller-based hardware.

The transformation matrices are shown in Equations 8.4 and 8.5:

$$\begin{bmatrix} X_\alpha \\ X_\beta \\ X_0 \end{bmatrix} = \frac{2}{3} \begin{bmatrix} 1 & \dfrac{-1}{2} & \dfrac{-1}{2} \\ 0 & \dfrac{-\sqrt{3}}{2} & \dfrac{\sqrt{3}}{2} \\ \dfrac{1}{2} & \dfrac{1}{2} & \dfrac{1}{2} \end{bmatrix} \times \begin{bmatrix} X_a \\ X_b \\ X_c \end{bmatrix} \tag{8.4}$$

$$\begin{bmatrix} X_d \\ X_q \\ X_0 \end{bmatrix} = \begin{bmatrix} \cos(\theta) & \sin(\theta) & 0 \\ -\sin(\theta) & \cos(\theta) & 0 \\ 0 & 0 & 1 \end{bmatrix} \times \begin{bmatrix} X_\alpha \\ X_\beta \\ X_0 \end{bmatrix} \tag{8.5}$$

The real and reactive power generated by a DG can be controlled through current or voltage regulation, as depicted by Figure 8.13 and Equations 8.2–8.5.

As already discussed in this chapter, the set points for active (P_{ref}) and reactive power (Q_{ref}) are defined by the MPPT control for renewable energy resource, or they can be determined at an energy management supervisory level. The control diagram of the current can be simplified as depicted in Figure 8.14.

The d and q control loops have similar dynamics. Thus, tuning the PI current controller for one axis (say, the d-axis) will give same controller gains for the q-axis. The transfer function of the system is indicated in Equations 8.6–8.10. The PI controller commanding the current components in d–q frame will give reference voltages, which must be changed to abc-frame, so an SPWM can control the grid-connected inverter. The reader can study and replace the SPWM by a space vector modulation. Figure 8.15 shows the full SPWM SVI. Figures 8.16 and 8.17 show the

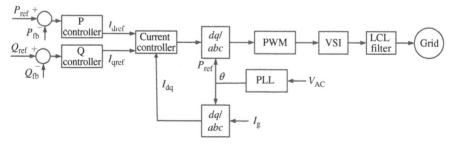

FIGURE 8.13 Control of the grid-connected inverter.

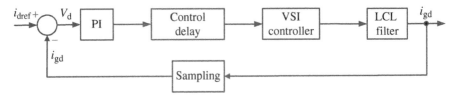

FIGURE 8.14 Simplified current control loop.

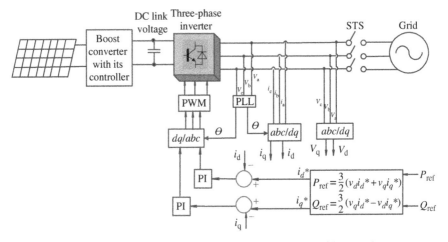

FIGURE 8.15 Grid-connected photovoltaic power system with control system.

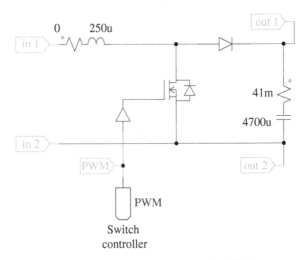

FIGURE 8.16 Boost converter in PSIM.

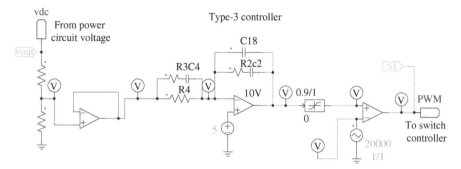

FIGURE 8.17   Implemented type III controller for boost converter in PSIM.

boost converter and its controller in PSIM environment. Figure 8.18 shows the whole grid-connected inverter system model in PSIM:

$$G_{PI}(s) = K_p + \frac{K_i}{s} \tag{8.6}$$

$$G_{control}(s) = \frac{1}{1+sT_s} \tag{8.7}$$

$$G_{sampling}(s) = \frac{1}{1+0.5sT_s} \tag{8.8}$$

$$G_{inverter}(s) = \frac{1}{1+0.5sT_{sw}} \tag{8.9}$$

$$G_{cl}(s) = G_{PI} * G_{control} * G_{inverter} * G_{filter} * G_{sampling} \tag{8.10}$$

Figure 8.19 shows the voltages V_1, V_2, and V_3 which are the control signals in grid-connected mode and the voltages V_4, V_5, and V_6 which are the control signals in islanded mode. As shown in Figure 8.19, the multiplexer selects either V_1, V_2, V_3 in grid-connected mode or V_4, V_5, V_6 in islanded mode, and depending on the mode of the system, which applies one of those sets to the inverter, since that is basically a multiplexed switch. Figure 8.20 shows the sinusoidal pulse-width modulator in PSIM that commands the modulating signals.

Some results are portrayed in Figures 8.21, 8.22, 8.23, 8.24, and 8.25. The reference value for the active power is 2 kW and for the reactive power is 1 kVAR initially. At $t=0.5$ the reference values for active and reactive power change to 3 kW and 1.5 kVAR, respectively. At the beginning the inverter is in grid-connected mode, so the main purpose of the inverter is controlling current and injecting active and reactive power to the grid. Figures 8.21 and 8.22 show the reference and actual values for active and reactive powers, respectively.

In order to show a good controller performance for boost converter for controlling the DC link voltage, the input voltage is changed at $t=0.6$ s from 95 to 80 V. Figure 8.23

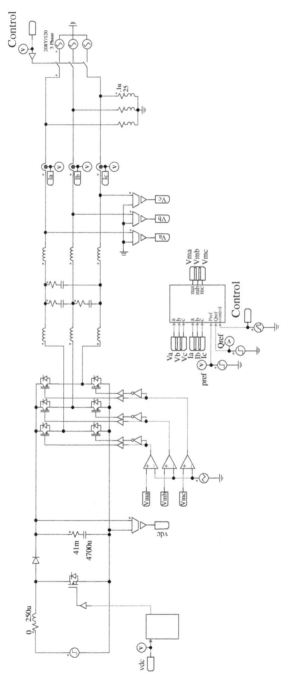

FIGURE 8.18 Grid-connected PV system in PSIM.

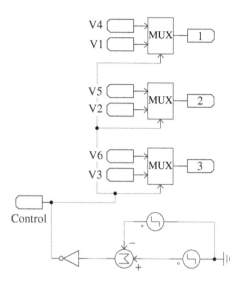

FIGURE 8.19 Implemented circuit for multiplex controller in PSIM for transition of grid-connected to islanded modes.

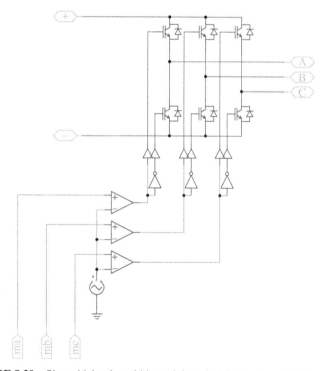

FIGURE 8.20 Sinusoidal pulse width modulator in PSIM and modulating signals.

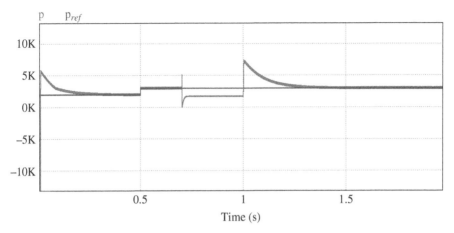

FIGURE 8.21 Reference and actual values for active power.

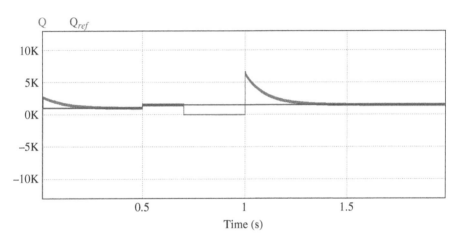

FIGURE 8.22 Reference and actual values for active power.

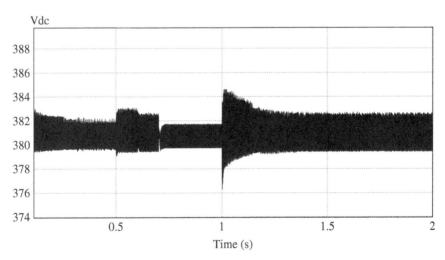

FIGURE 8.23 DC link voltage.

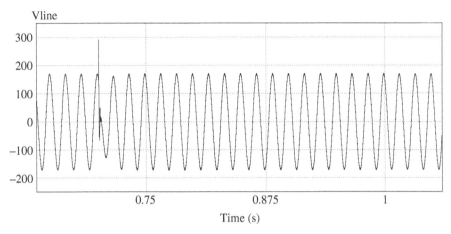

FIGURE 8.24   The voltage at the PCC.

shows the DC link voltage, and it is clear from the figure that it is fairly stable with changing the input voltage.

At time $t = 0.7$ s to $t = 1$ s the inverter goes to islanded mode, and because the main purpose of the controller in this mode is to control the PCC voltage, inverter does not control active and reactive power during this period, and it is shown in Figures 8.21 and 8.22. Figure 8.24 shows the PCC voltage; as shown in this figure, the voltage in PCC has same magnitude and frequency with the main grid. Figure 8.25 shows the output current of the inverter. At $t = 1$ s the inverter goes back to the grid-connected mode and the controller switches from voltage to current control mode.

8.9   SUGGESTED PROBLEMS

1. Study how a peak power tracking control can be implemented for a PV solar energy system. Design a controller for grid-connected PV inverter with capability of maximum power point tracking. Write your peak power tracking algorithm inside of a simplified C-block in PSIM.

2. Repeat the same design of a peak power tracking for a grid-connected PV inverter in Simulink. Write your peak power tracking algorithm inside of a Matlab called function.

3. Study how a peak power tracking control can be implemented for a wind energy system. Design a controller for grid-connected wind inverter with capability of maximum power point tracking. Write your peak power tracking algorithm inside of a simplified C-block in PSIM.

4. Repeat the same design of a peak power tracking for a grid-connected wind inverter in Simulink. Write your peak power tracking algorithm inside of a Matlab called function.

(a)

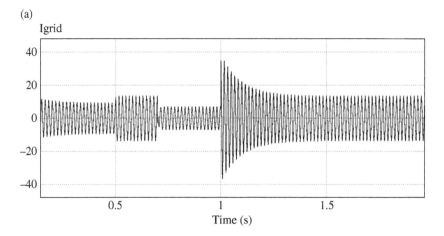

(b)

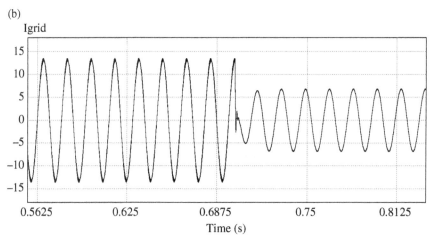

(c)

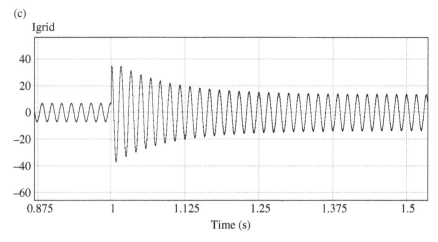

FIGURE 8.25 Inverter current response for connection and disconnection from the grid. (a) Output current of inverter in a long time span, (b) output current of inverter when it goes from grid connected to islanding at $t=0.7$ s, and (c) output current of inverter when it goes from islanding to grid connected at $t=1$ s.

5. Study PLLs for grid synchronization. Design a PLL capable of tracking the positive sequence and the negative sequence of a three-phase imbalanced utility grid.

6. Study how a single-phase inverter is implemented. Design in PSIM and in Matlab and compare the methodologies.

7. An LCL filter is very important for grid-connected inverters. Design one LCL filter and use it in your projects above.

8. Study how an inverter can be controlled to transition from a grid-connected control and automatically transfer to islanded conditions after disturbances are monitored on the utility voltage or a blackout of the utility voltage is detected. Your system should be able to maintain some local loads always on during the transition and should operate in accordance to IEEE 1547 and IEEE 519 standards.

9. Study how a three-phase inverter can be used for harmonic minimization using (i) a feedback connection, (ii) a feedforward connection, (iii) passive filtering, and (iv) active filtering. Implement your simulation in both PSIM and Matlab/ Simulink.

REFERENCES

[1] YANG, Y., BLAABJERG, F., WANG, H. and SIMÕES, M.G., "Power control flexibilities for grid-connected multi-functional photovoltaic inverters", in Proceedings of the 4th International Workshop on Integration of Solar Power into Power Systems, Berlin, IET, pp. 233–239, November 10–11, 2014.

[2] MAKNOUNINEJAD, A., SIMÕES, M.G. and ZOLOT, M., "Single phase and three phase P+Resonant based grid connected inverters with reactive power and harmonic compensation capabilities", IEEE International Electric Machines and Drives Conference, 2009. IEMDC '09, Miami, IEEE, pp. 385–391, May 3–6, 2009.

[3] CHAKRABORTY, S., SIMÕES, M.G. and KRAMER, W., Power Electronics for Renewable and Distributed Energy Systems: A Sourcebook of Topologies, Control and Integration, Springer-Verlag, New York, ISBN-10: 1447151038, ISBN-13: 978-1447151036 (Citations – GoS: 3), 2013.

[4] BROD, D.M. and NOVOTNY, D.W., "Current control of VSI-PWM inverters", IEEE Transactions on Industry Applications, vol. IA-21, pp. 562–570, 1985.

[5] HABETLER, T.G., "A space vector-based rectifier regulator for ac/dc/ac converters", IEEE Transactions on Power Electronics, vol. 8, pp. 30–36, 1993.

[6] VAN DER BROECK, H.W., SKUDELNY, H.C. and STANKE, G., "Analysis and realization of a pulse width modulator based on space vector theory", IEEE Transactions on Industry Applications, vol. IA-24, pp. 142–150, 1988.

[7] CARNIELETTO, R., BRANDÃO, D.I., SURYANARAYANAN, S., FARRET, F.A. and SIMÕES, M.G., "Smart grid initiative", IEEE Industry Applications Magazine, vol. 17, no. 5, pp. 27–35, 10.1109/MIAS.2010.939651, September/October 2011.

[8] SUKEGAWA, T., KAMIYAMA, K., TAKAHASHI, J., IKIMI, T. and MATSUTAKE, M., "A multiple PWM GTO line-side converter for unity power factor and reduced harmonics", IEEE Transactions on Industry Applications, vol. 28, no. 6, pp. 1302–1308, 10.1109/28.175281, 1992

[9] AKAGI, H., KANAZAWA, Y., FUJITA, K. and NABAE, A., "Generalized theory of instantaneous reactive power and its application", Electrical Engineering in Japan, vol. 103, pp. 58–66, 10.1002/eej.4391030409, 1983.

[10] OHNISHI, T., "Three-phase PWM converter/inverter by means of instantaneous active and reactive power control", Proceedings. IECON '91, 1991 International Conference on Industrial Electronics, Control and Instrumentation, Kobe, IEEE, pp. 819–824, vol. 1, October 28 to November 1, 1991, doi:10.1109/IECON.1991.239183.

[11] IEEE Standards Coordinating Committee 21 on Fuel Cells, Photovoltaics, Dispersed Generation, and Energy Storage, Institute of Electrical and Electronics Engineers, IEEE-SA Standards Board, IEEE Guide for Conducting Distribution Impact Studies for Distributed Resource Interconnection, Institute of Electrical and Electronics Engineers, New York, IEEE Std 1547.7-2013, doi:10.1109/IEEESTD.2014.6748837, pp. 1–137, 2014.

[12] Inverters, Converters, Controllers and Interconnection System Equipment for Use with Distributed Energy Resources UL1741. Underwriters Laboratory (UL).

[13] YAZDANI, A. and IRAVANI, R., Voltage-Sourced Converters in Power Systems: Modeling, Control, and Applications, John Wiley & Sons, Inc., New York, ISBN:978-0-470-52156-4, 2010.

[14] SHEN, G., ZHU, X., CHEN, M. and XU, D., "A new current feedback PR control strategy for grid-connected VSI with an LCL filter", Twenty-Fourth Annual IEEE Applied Power Electronics Conference and Exposition (APEC), Washington, DC, IEEE, pp. 1564–1569, February 15–19, 2009.

[15] FUKUDA, S. and YODA, T., "A novel current-tracking method for active filters based on a sinusoidal internal model for PWM inverters", IEEE Transactions on Industry Applications, vol. 37, no. 3, pp. 888–895, 2001.

[16] DASH, A.R., BABU, B.C., MOHANTY, K.B. and DUBEY, R., "Analysis of PI and PR controllers for distributed power generation system under unbalanced grid faults", International Conference on Power and Energy Systems, Chennai, IEEE, pp. 1–6, December 22–24, 2011.

[17] CUIYAN, L., DONGCHUN, Z. and XIANYI, Z., "A survey of repetitive control", Proceedings of 2004 IEEE/RSJ International Conference on Intelligent Robots and Systems (IROS 2004), IEEE, pp. 1160–1166, vol. 2, doi:10.1109/IROS.2004.1389553, September 28 to October 2, 2004.

[18] ROSHAN, A., A DQ rotating frame controller for single-phase full-bridge inverter used in small distributed generation systems, Master Thesis, Virginia Polytechnic Institute, 2006.

[19] TEODORESCU, R., BLAABJERG, F., LISERRE, M. and LOH, P.C., "Proportional-resonant controllers and filters for grid-connected voltage-source converters", IEE Proceedings Electric Power Applications, vol. 153, no. 5, pp. 750–762, September 2006.

[20] TEODORESCU, R., LISERRE, M. and RODRIGUEZ, P., Grid Converters for Photovoltaic and Wind Power Systems, John Wiley & Sons, Ltd, Chichester, 2011, p. 416.

[21] MEERSMAN, B., DE KOONING, J., VANDOORN, T., DEGROOTE, L., RENDERS, B. and VANDEVELDE, L., "Overview of PLL methods for distributed generation units", Proceedings of the 45th International Universities Power Engineering Conference (UPEC), Cardiff, IEEE, pp. 1–6, August 31 to September 3, 2010.

[22] KAURA, V. and BLASKO, V., "Operation of a phase locked loop system under distorted utility conditions", IEEE Transactions on Industry Applications, vol. 33, no. 1, pp. 58–63, 1997.

[23] MARAFÃO, F.P., DECKMANN, S.M., POMILIO, J.A. and MACHADO, R.Q., "A software-based PLL model: analysis and applications," In Brazilian Automatic Conference (CBA), Brazilian Automatic Society (SBA—Sociedade Brasileira de Automação), Convention Center of Gramado, Rio Grande do Sul, Brazil, September 21–24, 2004.

[24] LISERRE, M., BLAABJERG, F. and HANSEN, S., "Design and control of an LCL-filter-based three-phase active rectifier", IEEE Transactions on Industry Applications, vol. 41, no. 5, pp. 1281–1291, 10.1109/TIA.2005.853373, 2005.

[25] REZNIK, A., SIMÕES, M.G., AL-DURRA, A. and MUYEEN, S.M., "LCL filter design and performance analysis for grid interconnected systems", IEEE Transaction on Industry Applications, vol. 50, no. 2, pp. 1225–1232, 10.1109/TIA.2013.2274612, March–April 2014.

FURTHER READING

DUARTE, J.L., VAN ZWAM, A., WIJNANDS, C. and VANDENPUT, A., "Reference frames fit for controlling PWM rectifiers", IEEE Transactions on Industrial Electronics, vol. 46, no. 3, pp. 628–630, 10.1109/41.767071, 1999.

HARIRCHI, F., SIMÕES, M.G., AL-DURRA, A. and MUYEEN, S.M., "Short transient recovery of low voltage grid tied DC distributed generation", IEEE Energy Conversion Congress and Exposition (ECCE), Montreal, QC, IEEE, September 20–24, 2015.

HARIRCHI, F., SIMÕES, M.G., BABAKMEHR, M., AL-DURRA, A. and MUYEEN, S.M., "Designing smart inverter with unified controller and smooth transition between grid-connected and islanding modes for microgrid application", IEEE Industry Applications Society Annual Meeting (IAS), Dallas, TX, IEEE, pp. 18–22, October 18–22, 2015.

KROPOSKI, B., PINK, C., DEBLASIO, R., THOMAS, H., SIMÕES, M.G. and SEN, P.K., "Benefits of power electronic interfaces for distributed energy systems", IEEE Transactions on Energy Conversion, vol. 25, no. 3, pp. 901–908. 10.1109/TEC.2010.2053975, 2010.

LINDGREN, M.B., "Feedforward-time efficient control of a voltage source converter connected to the grid by low-pass filters", Proceedings of the 26th Annual IEEE Power Electronics Specialists Conference, 1995. PESC '95 Record, Atlanta, IEEE, doi:10.1109/PESC.1995.474942, June 18–22, 1995.

LUTE, C.D., SIMÕES, M.G., BRANDÃO, D.I., AL-DURRA, A. and MUYEEN, S.M., "Experimental evaluation of an interleaved boost topology optimized for peak power tracking control", Proceedings of the 40th Annual Conference of the IEEE Industrial Electronics Society (IECON), Dallas, TX, IEEE, October 29 to November 1, 2014.

MALINOWSKI, M., KAZMIERKOWSKI, M.P. and TRZYNADLOWSKI, A.M., "A comparative study of control techniques for PWM rectifiers in AC adjustable speed drives",

IEEE Transactions on Power Electronics, vol. 18, no. 6, pp. 1390–1396, 10.1109/TPEL.2003.818871, 2003.

MAO, J.F., WU, G.Q., WU, A.H., ZHANG, X.D. and YANG, K., "Modeling and decoupling control of grid-connected voltage source inverter for wind energy applications", Advanced Materials Research, vol. 213, pp. 369–373, 10.4028/www.scientific.net/AMR.213.369, 2011.

MIRANDA, U.A., AREDES, M. and ROLIM, L.G.B., "A DQ synchronous reference frame current control for single-phase converters", IEEE Power Electronics Specialists Conference, vol. 2, no. Conf 36, pp. 1377–1381, 10.1109/PESC.2005.1581809, 2005.

YU, X.Y., CECATI, C., DILLON, T. and SIMÕES, M.G., "The new frontier of smart grids", IEEE Industrial Electronics Magazine, vol. 5, no. 3, pp. 49–63, 10.1109/MIE.2011.942176, November/December 2011.

9

MODELING ALTERNATIVE SOURCES OF ENERGY

9.1 ELECTRICAL MODELING OF ALTERNATIVE POWER PLANTS

A detailed simulation of small power plants may be a very cost-effective solution, and very often several subsystems of a power plant might be inappropriate, difficult to find, or very expensive, like photovoltaic (PV) panels and fuel cells. The student, designer, or engineer should know how such power plants can be formulated at least with some approximated models in order to study their control and automation or their best operation. Furthermore, it is advisable to understand how these power plants will behave in cases of emergency, ride-through, short circuits, standby, general surges, and operations close to the limits of the overall electrical generation and distribution.

In this chapter, it discussed and simulated the most usual small electrical power plants such as wind, photovoltaic, fuel cells, and primary movers associated with batteries (the most common electrical storage). For the simulations proposed in this chapter, the authors considered well-known models found in the current literature. However, all those models can be modified for particular applications, or more detailed functions, to serve as case studies for distributed and sparse generation [1, 2].

Modeling Power Electronics and Interfacing Energy Conversion Systems, First Edition.
M. Godoy Simões and Felix A. Farret.
© 2017 John Wiley & Sons, Inc. Published 2017 by John Wiley & Sons, Inc.

9.2 MODELING A PHOTOVOLTAIC POWER PLANT

A well-known model of a PV cell is based on the *one-diode model*, presented in Figure 9.1. A brief mathematical description follows next [1, 3].

Based on Figure 9.1, current I_L can be calculated as indicated in Equation 9.1 [1]:

$$I_L = \frac{R_p}{R_p + R_s + R_L}\left\{ I_g - I_s\left(e^{qV_L(1+R_s/R_L)/kT} - 1\right)\right\} \tag{9.1}$$

Figure 9.2 illustrates the PSIM PV cell model with the parameters listed in Table 9.1. A voltmeter, an ammeter, and a wattmeter are connected to the terminals of the load resistance to allow the plot of its output characteristics. In this figure, these instruments are set up by function blocks though PSIM instruments can do the same job. This configuration is arranged in such a way that the output current and power are swept by a triangular wave changing gradually the load resistance, R_L. The output characteristics of voltage and power are plotted in Figure 9.3.

A complete PV panel is made of many cells connected in series–parallel arrangements to obtain higher electrical values. The output voltage can be regulated according to the user needs to a lower level using, for example, a buck

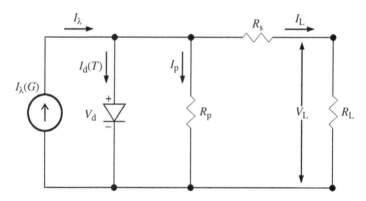

FIGURE 9.1   One-diode model for a photovoltaic (PV) cell.

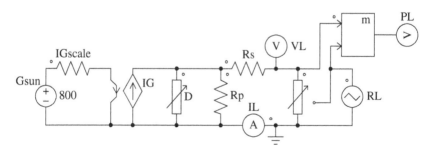

FIGURE 9.2   PSIM one-diode model for a photovoltaic cell.

TABLE 9.1   Parameters of the PV Cell

Parameter	Magnitude
Maximum power	2.86 W
Open-circuit voltage	1.00 V
Short-circuit current	4.00 A
Voltage across the load	0.78 V
Current through the load	3.67 A
R_s	0.03 Ω
R_p	200 Ω
R_L	5 Ω
G	800 W/m²
T	298.15 K
I_s	100×10^{-12} A
η	1.57

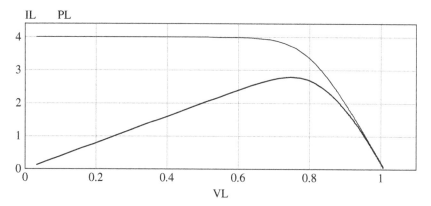

FIGURE 9.3   Power and voltage characteristics for a single PV cell.

converter, or to a higher level using a boost converter. It is interesting to note in Figure 9.3 the current–source behavior of a single PV cell, making clear that this is typically a current source. However, when PV cells are connected in series, there is a large increment in the series resistance which causes a remarkable negative slope in such current–source characteristic, as it is observed in all commercial catalogs [1].

9.3   MODELING AN INDUCTION GENERATOR (IG)

The usual per-phase model of an induction machine operating either as generator $(n > n_s)$ or motor $(n < n_s)$ is presented in Figure 9.4 with parameters listed in Table 9.2. For example, for the four-pole machine, the synchronous speed n_s is 1800 rpm. A detailed mathematical description of this model is presented in Ref. [1].

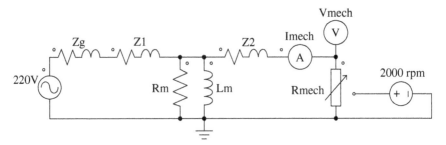

FIGURE 9.4   Per-phase model of an induction generator connected to grid.

TABLE 9.2   SEIG Parameters

Parameter	Magnitude
Rated voltage	220 V
Rated current	6 A
Rotor speed	2000 rpm
p	4 poles
R_g	10.53 Ω
L_g	57.4 mH
R_1	0.542 Ω
R_2	0.427 Ω
L_1	1.14 mH
L_2	1.4 mH
R_m	114 Ω
L_m	127 mH

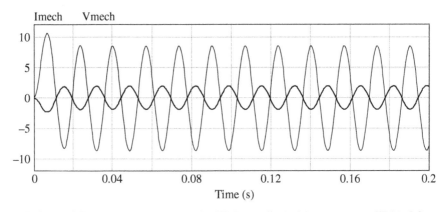

FIGURE 9.5   Simulation results of the SEIG described with parameters of Table 9.2.

The equivalent mechanical resistance including the s slip factor changes is inserted as a PSIM rpm-dependent nonlinear resistance defined as

$$R_{mech} = \frac{R_2\left(1-s\right)}{s} = \frac{R_2}{\left(n_s / n\right) - 1}$$

The current through the nonlinear element gives the polarity to variable across it.

For a three-phase machine, it is usually sufficient to consider one equivalent circuit per phase since the manufacturer makes sure their machine is a well-balanced and high-quality product. So, the steady-state features such as power factor, losses, and general performance can be established under a per-phase basis. For advanced control methods, the reader is suggested to study [1] which covers scalar and vector control of d–q induction machines and generators. In the output voltage simulation of Figure 9.5, the opposition of phase between voltage and current characterizing power generation in respect to the source is noticeable.

9.4   MODELING A SEIG WIND POWER PLANT

The SEIG modeled in Figure 9.6 is associated to a wind power turbine with parameters listed in Table 9.2. The reason why a 1 nF capacitor is included in this model was to simulate the residual magnetism of the induction generator necessary to start the self-excitation process. A brief mathematical description of this model is presented in Ref. [1]. Similarly, to PV energy systems, the output voltage of the induction generator can be regulated to a lower level using a buck converter or to

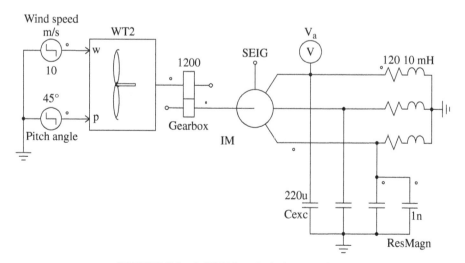

FIGURE 9.6   A SEIG-based wind power plant.

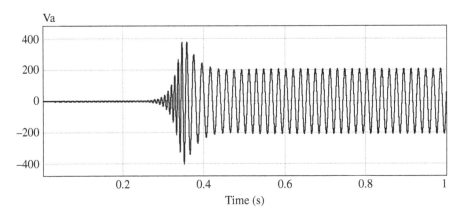

FIGURE 9.7   Self-excitation output voltage of a SEIG connected to a wind turbine.

a higher level with a boost converter. The output voltage during the self-excitation process is illustrated in Figure 9.7.

9.5   MODELING A DFIG WIND POWER PLANT

In this section, a doubly fed induction generator (DFIG) is mechanically connected to a wind turbine as indicated in Figure 9.8. A brief mathematical description of this model is presented in Ref. [1]. Notice that the three-phase generator connected to the rotor winding used in this model can be replaced by a power converter and easily adjusted to obtain several different features out of this configuration. The wind power plant model illustrated in Figure 9.8 used the parameters listed in Table 9.3.

Figure 9.9 shows the response of the rotor voltage and stator current for the wind turbine-powered system. A low inrush stator current in this diagram is noticeable since it was not adopted any measure to avoid it.

9.6   MODELING A PMSG WIND POWER PLANT

A model of a permanent magnet synchronous generator (PMSG) mechanically connected to a wind power turbine is presented in Figure 9.10. A brief mathematical description of this model is encountered in Ref. [1]. The wind speed and blade pitch can be adjusted as well as the speed multiplication (gearbox ratio) to have the best results in terms of power generation and efficiency or to fit a particular purpose. The parameters necessary for this simulation are presented in Table 9.4 for the PMSG and load and Table 9.5 for the wind turbine.

Figure 9.11 illustrates the start-up phase output voltage of a PMSG connected to a wind turbine. This output voltage can be regulated to a lower level using, for example, a buck converter or to a higher level with a boost converter.

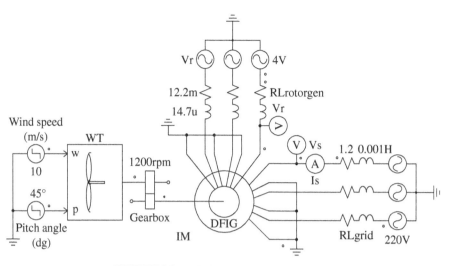

FIGURE 9.8 DFIG-based wind power plant.

TABLE 9.3 Parameters of the DFIG

Parameter	Value
Rated voltage	220 V
Rated current	4 A
Rotor speed	1200 rpm
p	6 poles
R_{grid}	1.2 Ω
L_{grid}	1 mH
R_{rotgen}	12.2 Ω
L_{rotgen}	14.7 Ω
R_s	0.22 Ω
L_s	114 mH
R_r	127 mΩ
L_r	103 mH
R_m	127 Ω
L_m	114 mH

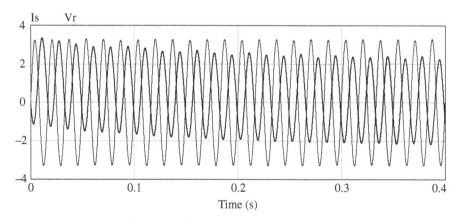

FIGURE 9.9 Rotor voltage and stator current for the DFIG system.

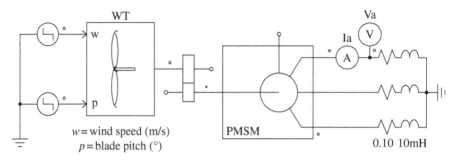

w = wind speed (m/s)
p = blade pitch (°)

PMSM

0.10 10mH

FIGURE 9.10 A PMSG wind power plant.

TABLE 9.4 Parameters of the PMSG

Parameter	Value
Rated voltage	265 V
Rated current	5 A
Rotor speed	1500 rpm
p	4 poles
R_L	10 Ω
L_L	100 mH
R_s	4.3 Ω
L_d	27 mH
L_q	67 mH
V_{pk}/k_{rpm}	98.67
Shaft time constant	10 s
Moment of inertia	0.00179

TABLE 9.5 Parameters of the Wind Turbine

Wind speed	7 m/s
Blade pitch	45°
WT nominal power	20 kW
Moment of inertia	8000 kg·m²

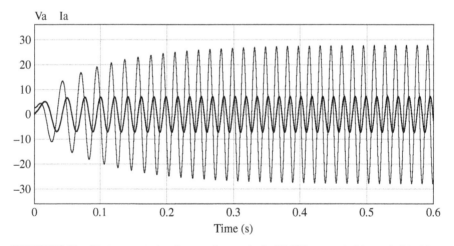

FIGURE 9.11 Start-up output voltage and current of a PMSG connected to a wind turbine.

9.7 MODELING A FUEL CELL STACK

Figure 9.12 shows the equivalent electric circuit of the electrochemical model of a proton exchange membrane (PEM) fuel cell (PEMFC) for which the KCL is implied by Equation 9.2:

$$E_{OC} - \frac{1}{C}\int i_C dt - (R_r + R_L) \cdot i_{fc} \tag{9.2}$$

The electrical equivalent model for the fuel cell (FC) activation R_{act}, concentration R_{con}, and ohmic R_r resistances is nonlinear. Therefore, the approach for solution should use numerical methods for nonlinear equations. Taking Equation 9.2 for $i_{fc} = i_a + i_c$ and the initial set of parameters, the mathematical calculations for a PEMFC stack have been implemented in the PSIM model indicated in Figure 9.13 [4].

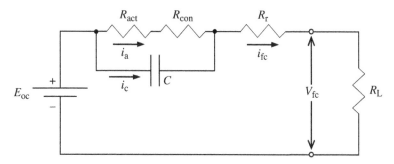

FIGURE 9.12 Electrochemical equivalent circuit of a PEMFC.

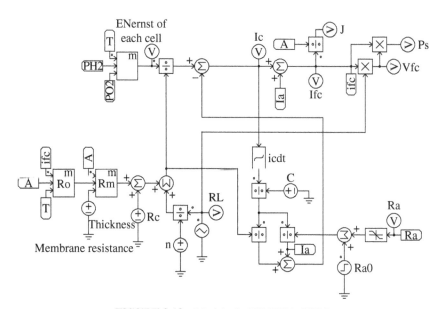

FIGURE 9.13 Model of a PEMFC in PSIM.

In order to determine the nonlinear resistance values including resistance activation and concentration, the Tafel equation is used to establish Equations 9.3 and 9.4:

$$R_a = R_{act} + R_{conc} = \frac{v_{act} + v_{conc}}{i_a} = \frac{v_c}{i_a} \tag{9.3}$$

$$i_a = \frac{v_c}{R_{act} + R_{con}} = \frac{1}{R_a C} \int i_c . dt \tag{9.4}$$

Combining Equations 9.2, 9.3, and 9.4 comes

$$E - \frac{1}{C} \int i_c . dt - (R_r + R_L) . \left(\frac{1}{R_a C} \int i_c . dt + i_c \right) = 0 \tag{9.5}$$

Dividing both sides of the identity (9.3) by $R_r + R_L$ and simplifying comes

$$\frac{E}{(R_r + R_L)} - \left[\frac{1}{(R_r + R_L) C} + \frac{1}{R_a C} \right] \int i_c . dt = i_c \tag{9.6}$$

The activation and concentration (mass transportation) resistances are defined, respectively, as

$$R_{act} = \frac{v_{act}}{i_{act}} = \frac{A}{i_{act}} . \ell n \left(\frac{J}{J_0} \right) \tag{9.7}$$

$$R_{conc} = \frac{v_{conc}}{i_{act}} = -\frac{B}{i_{act}} \ell n \left(1 - \frac{J}{J_{max}} \right) \tag{9.8}$$

The ohmic resistance of the cell membrane is defined by

$$R_m = \frac{\rho_M \ell}{A}$$

where

$$\rho_M = \frac{181.6 \left[1 + 0.03 (i_{FC} / A) + 0.062 (T / 303)^2 (i_{FC} / A)^{2.5} \right]}{\left[\psi - 0.634 - 3 (i_{FC} / A) \right] . e^{\left[4.18(T - 303)/T \right]}} \tag{9.9}$$

The activation voltage can be obtained from

$$V_{act} = -\left[\xi_1 + \xi_2 . T + \xi_3 . T . \ln \left(c_{O_2}^* \right) + \xi_4 . T . \ln \left(i_{FC} \right) \right]$$

where

$A_{\text{Tafel}} = RT/2\alpha F$, typically 0.06

$B = RT/4F = \alpha A$ for oxygen which is a constant that depends on the FC and its operating state

α is the charge transfer coefficient

$F = 96\,485.34\,\text{C/mol}$ is the Faraday's constant

$R = 8.314\,\text{J/K}\cdot\text{mol}$ t is the molar gas constant

$J = J_0 e^{2\alpha F\Delta V_{\text{act}}/RT}$ is the Butler–Volmer equation

J_0 is the current density at which the overvoltage begins to move from zero (Larminie, p. 49, 3rd line)

The charge transfer coefficient α ranges from 0 to 1.0 and is the proportion of the electrical energy applied that is harnessed in changing the rate of an electrochemical reaction. For a great variety of electrode materials and in particular for the hydrogen electrode, $\alpha = 0.5$. Therefore,

$$R_{\text{a}} = \frac{A}{i_{\text{act}}}\cdot \ell n\left(\frac{J}{J_0}\right) - \frac{B}{i_{\text{act}}}\ell n\left(1 - \frac{J}{J_{\text{max}}}\right) \qquad (9.10)$$

For the simulation it can be assumed that the FC is comprised of n identical cells to form a stack, and Equation 9.6 becomes

$$\frac{n\cdot E}{\left(R_{\text{L}} + n\cdot R_{\text{r}}\right)} - \frac{1}{\left(R_{\text{L}} + n\cdot R_{\text{r}}\right)\cdot(C/n)}\int i_{\text{c}}.dt - \frac{1}{n\cdot R_{\text{a}}\cdot(C/n)}\int i_{\text{c}}.dt = i_{\text{c}} \qquad (9.11)$$

With some algebraic manipulation, the equation can be adapted for simulation purposes as shown in Equation 9.12:

$$\frac{E}{\left(\dfrac{R_{\text{L}}}{n} + R_{\text{r}}\right)} - \left[\frac{1}{\left(\dfrac{R_{\text{L}}}{n} + R_{\text{r}}\right)} + \frac{1}{R_{\text{a}}}\right]\frac{\int i_{\text{c}}.dt}{C} = i_{\text{c}} \qquad (9.12)$$

A PEMFC called *Ballard Mark V* can be simulated in accordance with the diagram in Figure 9.12 with the equations described in this section and parameters listed in Table 9.6. In this table it used the definitions of oxygen and hydrogen concentration as given by Equations 9.13 and 9.14, where J_n is the current density equivalent to the internal currents and direct fuel passage (A/cm²) [5–7]:

$$c^*_{O_2} = \frac{P_{O_2}}{5.08\cdot 10^6\,e^{-498/T}} \qquad (9.13)$$

$$c^*_{H_2} = \frac{P_{H_2}}{1.09\cdot 10^6\,e^{77/T}} \qquad (9.14)$$

TABLE 9.6 Typical Parameters of a Ballard Mark V Fuel Cell

Parameter	Value	Parameter	Value
T	343.15 K	ξ_1	-0.948
A	50.6 cm²	ξ_2	$0.00286 + 0.0002 \ln A + (4.3 \times 10^{-5}) \ln C_{H_2}^*$
ℓ	178 μm	ξ_3	7.6×10^{-5}
$c_{O_2}^*$	1.10⁻⁴ mol/cm³	ξ_4	-1.93×10^{-4}
$c_{H_2}^*$	1.10⁻⁴ mol/cm³	ψ	23.0
$p_{O_2}^*$	1.0 atm	J_{max}	1.5 A/cm²
$p_{H_2}^*$	1.3 atm	J_n	2 mA/cm²
RC	0.001 Ω	n	32
A_{Tafel}	0.03	α	0.5
B	0.015 V	C	3 F

The expression (9.15) is used to determine for the open-circuit voltage E_{oc} [1]:

$$E_{oc} = E_{Nernst} = 1.229 - 0.85.10^{-3}(T - 298.15) + 4.31.10^{-5}T\left[\ln\left(p_{H_2}^*\right) + \frac{1}{2}\ln\left(p_{O_2}^*\right)\right]$$

(9.15)

where

R_L is the given load resistance
E_{oc} is the open-circuit voltage obtained by measurement
V_r is the voltage drop across R_r during the load interruption test
V_L, I_L are the load voltage and current, respectively, before the load interruption test

The circuit presented in Figure 9.12 with the subcircuit of Figure 9.14 was simulated in PSIM as shown in Figure 9.13. The nonlinear resistance in this model was simulated using the exchange current density parameters, initial current density, maximum current density, effective membrane area, H₂ and O₂ working pressures, and calculation blocks for v_{act} and v_{con} [1, 8–11].

During the FC simulations it is important to take into account that:

1. $J < J_{max}$ to avoid logarithms of zero or negative numbers in the definition of the stress concentration.
2. The slope of the resistance (voltage) charge should always be less than the total simulation time, or:

$$f < 1/t_{simulation}$$

(9.16)

3. The value of the capacitance C should be evaluated by numerical method.

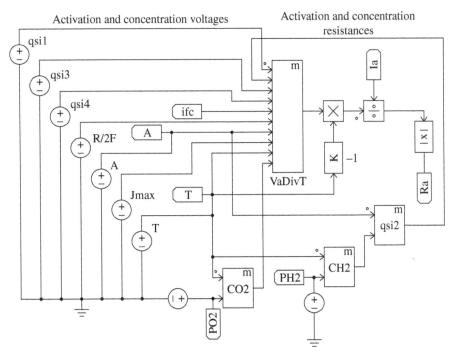

FIGURE 9.14   Subcircuit of the nonlinear resistance R_a.

9.8   MODELING A LEAD ACID BATTERY BANK

Battery modeling is a quite complicated matter. The detailed representation is out of
the scope of this book. Description of equivalent circuits and parameters can be
found in several references [1, 12–14]. In this book the authors tried to discuss a reasonable and simple model such as the one presented in Figure 9.15. Determination of
charge and discharge rates and states of a battery can be made from a sequence of
tests. Reference [12] presents a more generic battery discharge/charge model that can
be used to describe the lead acid battery or any other battery electrochemistry.

In order to design integrated battery-based power electronics, a very good battery
behavioral model is required which may include several parameters like state of
charge (SOC), terminal voltage, cell temperature, internal pressure, and so on. There
are many battery models in the literature. Therefore, choosing a more complex or a
more simplified model depends on the application. In a typical electrochemistry-
based battery model, the applied current is the input, and the terminal voltage is the
output. However, for the model in this book, it was necessary to reverse the input–
output relationship of the current and voltage, because the modeling of integrated
battery systems requires a voltage source as the input while the system operates in
current-mode control in respect to the power flow to the utility grid. A decision
algorithm makes use of the SOC curve in order to prescribe the operation mode.

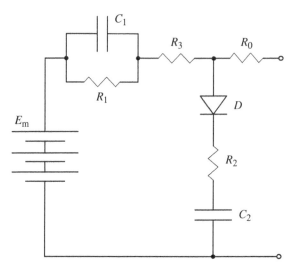

FIGURE 9.15   Battery model for a time-varying voltage.

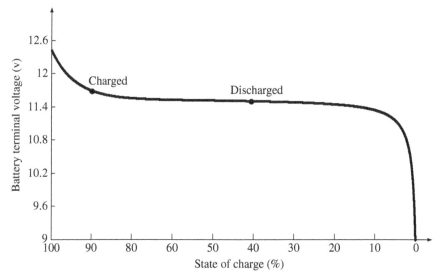

FIGURE 9.16   The SOC curve and the threshold points used in the decision algorithm.

Therefore, a third-order battery model is presented in this section, since it can guarantee the correct operation of an integrated battery system with a power electronic converter. Figure 9.16 presents a typical SOC curve, obtained through a simulation [15], with two threshold points. The threshold points can be used in a decision algorithm, and their locations on the SOC curve can be selected based on criteria of the spinning reserve. Figure 9.17 presents a third-order model of a lead acid battery,

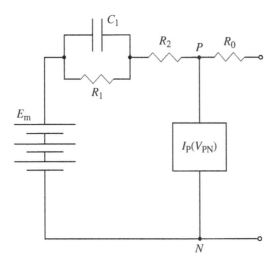

FIGURE 9.17   Third-order model of lead acid batteries.

where

   E_m is the open-circuit voltage
   C_1 is the overvoltage capacitance
   R_0 is the terminal resistance
   R_1 is the overvoltage resistance
   R_2 is the internal resistance
   $I_p(V_{PN})$ is the parasitic current in function of PN voltage, mainly caused by self-
      discharge (Equation 9.25)

   The model may be divided in three parts: (i) the main branch, composed of E_m, R_1, R_2, and C_1; (ii) the parasitic branch, composed of the block called IP(VPN); and (iii) the terminal branch, composed of the R_0. The battery's extracted charge is given by Equation 9.17, and the battery total capacity is given by Equation 9.18:

$$Q_{extr}(t) = Q_{extr\_ini} - \int_0^t i_m(\tau)d\tau \qquad (9.17)$$

where Q_{extr_ini} is the charge computed previously.

$$C(i,T_{emp}) = \frac{kC_0}{1+(k-1)(i/i_{nom})^\delta}\left(1 - \frac{T_{emp}}{T_{empf}}\right)^\epsilon \qquad (9.18)$$

where

k is the multiplier gain
C_0 is the no-load capacity at 0°C

T_{emp} is the electrolyte temperature in °C and T_{empf} is the final temperature

δ, ε are constants derived from experimental observations

i_{nom} is the battery nominal current

The SOC is given by Equation 9.19:

$$SOC = 1 - \frac{Q_{extr}}{C\left(i, T_{emp}\right)}$$

(9.19)

The depth of discharge (DOD) is given by Equation 9.20:

$$DOD = 1 - \frac{Q_{extr}}{C\left(i_{avg}, T_{emp}\right)}$$

(9.20)

The overvoltage resistance is given by Equation 9.21:

$$R_1 = -k_{R_1} \ln\left(DOD\right)$$

(9.21)

where k_{R_1} is a constant.

The overvoltage capacitance is given by Equation 9.22:

$$C_1 = \frac{\tau_1}{R_1}$$

(9.22)

where τ_1 is main branch constant time.

The internal resistance is given by Equation 9.23:

$$R_2 = k_2 \frac{e^{\left(k_{21}(1-SOC)\right)}}{1 + e^{\left(k_{22}\frac{i}{i_{nom}}\right)}}$$

(9.23)

The terminal resistance is given by Equation 9.24:

$$R_o = R_{oo}\left[1 + k_a\left(1 - SOC\right)\right]$$

(9.24)

where R_{oo} is the terminal resistance at $SOC = 1$.

The parasitic current is given by Equation 9.25:

$$I_P\left(V_{PN}\right) = V_{PN} k_{pn} \exp\left[\frac{V_{PN} / \left(\tau_{pn} + 1\right)}{k_p} + k_3\left(1 - \frac{T_{emp}}{T_{empf}}\right)\right]$$

(9.25)

where τ_n is the constant time of the parasitic branch.

The battery terminal voltage is given by Equation 9.26:

$$v_t = E_m + i_m Z_{eq} + i R_o$$

(9.26)

TABLE 9.7   Parameters of the Lead Acid and Lithium-Ion Batteries

Parameters\Type	Lead Acid	Li-ion
E_0 (V)	12.47	3.37
R (Ω)	0.04	0.01
K (V/Ah)	0.0470	0.0076
A (V)	0.830	0.264
B (per Ah)	125	26.55

where Z_{eq} is given by Equation 9.27:

$$Z_{eq} = \frac{R_1 + R_2 + sR_1R_2C_1}{sR_1C_1 + 1} \tag{9.27}$$

Currents i_m and i are related as Equation 9.28:

$$I_p = i_m - i \tag{9.28}$$

Although a state-space formulation leading to a transfer function expansion for the battery system can be developed, the authors preferred the equation-oriented model for a circuit-based simulation in PSIM. The battery capacity is found by looking to the manufacturer's datasheet, and the constants k presented in the previous equations are not easy to determine, but they can be found through specific experimental tests; such procedure is beyond the scope of this chapter, but readers can have further information in [16–19]. The battery model must be useful in understanding how the battery behaves. Based on the behavior, the design of a battery charging controller takes account the aspects that are relevant for the system design. In this chapter, the SOC curve is used during operation; the current point of the SOC curve is estimated in order to be used as input information for a decision maker that is integrated with the energy management control of the renewable energy source used in the grid-connected or stand-alone system. Typical parameters for lead acid and lithium-ion batteries are depicted in Table 9.7.

9.9   MODELING AN INTEGRATED POWER PLANT

The models presented in this chapter allow further modeling studies or their use to establish an integrated power plant by aggregating the model of every source (source subroutine) and their own control device (control subroutine) to a general power bus (load control subroutine) [1, 20]. Other sources can be easily aggregated each one with their own parameters to form a general power plant. Such a model would look like the one presented in Figure 9.18. The subroutines of this layout are shown in Figures 9.19, 9.20, 9.21, 9.22, 9.23, 9.24, and 9.25. The plant subroutine parameters used in this figure are displayed in Tables 9.8, 9.9, 9.10, 9.11, and 9.12.

    Figure 9.26 displays the currents through the primary and secondary loads. The primary load is the one to be supplied with the maximum priority at an acceptable voltage

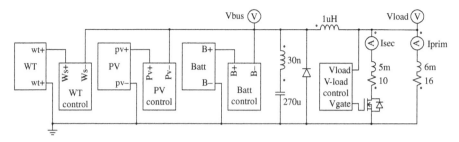

FIGURE 9.18 An integrated power plant.

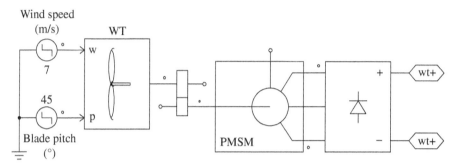

FIGURE 9.19 Subcircuit of the wind power generator.

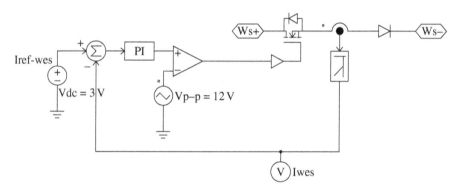

FIGURE 9.20 Wind power generator control.

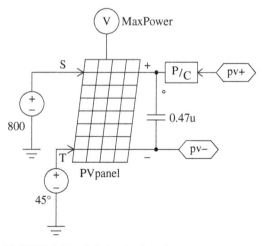

FIGURE 9.21 Subcircuit of the PV power generator.

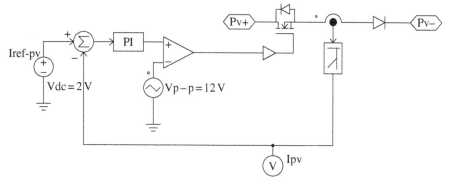

FIGURE 9.22 Control scheme of the PV power generator.

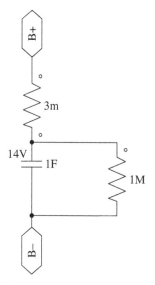

FIGURE 9.23 Subcircuit of a simple battery energy storage.

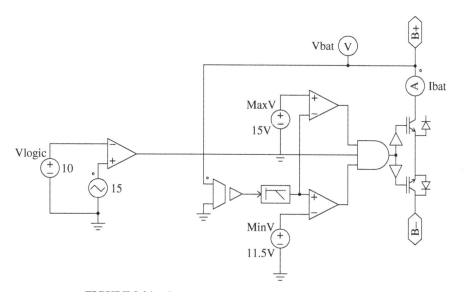

FIGURE 9.24 Control scheme of the battery energy storage.

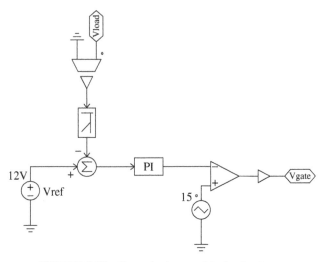

FIGURE 9.25 Control scheme of the load voltage.

TABLE 9.8   Parameters of the Integrated Power Plant

Parameter	Value
Primary load (R, L)	16 Ω, 6 mH
Secondary load (R, L)	10 Ω, 5 mH
Carrier frequency	15 kHz
Carrier magnitude	15 V
Control time constant	0.001 s
Control gain	1
Control reference voltage	12 V
Cutoff frequency	1000 Hz
Damping ration	0.8

TABLE 9.9   Parameters of the Wind Turbine

Wind speed	7 m/s
Blade pitch	45°
WT nominal power	20 kW
Moment of inertia	8000 kg·m²

TABLE 9.10   Parameters of the PMSG

Parameter	Magnitude
Rated power	10 kW
Rated wind speed	7 m/s
Rotor speed	1500 rpm
Moment of inertia	8000 kg·m²

TABLE 9.11   Parameters of the PV Panel

Parameter	Magnitude
Number of cells	36 cells
Light intensity (G)	800 W/m²
Reference temperature	42°C
Temperature coefficient	0.0024 A/K
Short-circuit current	3.8 A
Band energy	1.12 eV
Current through the load	3.67 A
Diode saturation current (I_s)	2.16×10^{-8} A
R_s of each cell	0.008 Ω
R_p of each cell	1000 Ω
Ideality factor (η)	1.12

TABLE 9.12   Parameters of the Lead Acid Battery

Parameters	Value
E_o	14 V
Stray shunt resistance	1 MΩ
Internal resistance	3 mΩ
Internal capacitance	1 F

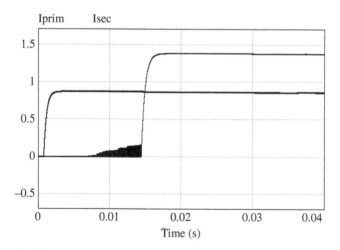

FIGURE 9.26   Currents through the primary and secondary loads.

level. The secondary load is the one using only the excess of generated power not in use by the primary load, such as battery charging, refrigeration, heating, water pumping, electrolysis, and irrigation. The secondary load does not need an uninterrupted supply.

In this integrated systems, a hill climbing control (HCC) sets up the electrical current of each individual power source [1]. The common load voltage is controlled by the secondary load current. The secondary load current gets all of the power generation so as to maximize the converted power even if any particular power source is not at their maximum power point. This is typical for alternative energy sources because if the primary source is able to produce more power than the main load consumes, it would not be converted, for example, excess of wind, water, or sunshine. The layout of sources and respective controls used in the model shown in Figure 9.18 can be easily extended to any number of sources and modified for several other possible renewable/alternative energy sources and their converters plus storage. Very complex economic-based models with real-time tariffs can be implemented plus random availability of wind and solar energy, with management of the SOC of batteries and power quality concerns at the point of common coupling (PCC).

9.10   SUGGESTED PROBLEMS

1. Develop a PSIM-based photovoltaic panel of six strings with 80 identical cells connected in parallel. Find their characteristics through the parameters listed in Table 9.11 in this chapter and calculate the total ohmic losses. What is the voltage regulation of this panel?

2. Design in PSIM a small wind power plant to feed an AC load with the following parameters: wind speed $= 7\,m/s$, blade pitch $= 40°$, $R_L = 5\,\Omega$, and $X_L = 25\,\Omega$. Simulate your system and show with results and analysis the characteristics $I_L \times V_L$ and $P_L \times V_L$ discussing how to reach the maximum power point (MPP) and how a controller could be implemented for optimization of the power conversion.

3. Do problem #2 again, but using Matlab/Simulink and the Power Systems Toolbox.

4. Develop the model for a fuel cell stack with 32 standard cells described in Table 9.6 in the integrated power plant shown in Figure 9.17 and observe how a battery can have a better cycling in their operation with such hybrid system.

5. Search in literature the typical parameters for NiMH and NiCd batteries making their PSIM simulation for charge and discharge states at rated load.

6. Develop a battery model to reproduce as close as possible the output characteristic analog to the model shown in Figure 9.25. How long will this battery take to discharge 50% of their capacity? Connecting this battery to a $5\,\Omega$ resistor, what is the battery output voltage? When charging and discharging, what is the voltage variation across the battery?

7. A third-order battery model is described in Section 9.7; further details can be found on Refs. [16–19]. Implement a PSIM and a Matlab-based third-order battery model, compare with the one developed on problem #6, and explore the use of a more complex model when compared to a simpler one.

8. Suggest a sequence of tests to obtain the following parameters of a 12V battery: R_0, R_1, R_2, R_3, C_1, and C_2. Implement a simulation model to emulate such test sequence.

REFERENCES

[1] FARRET, F.A. and SIMÕES, M.G., Integration of Alternative Sources of Energy, IEEE-Wiley Interscience, John Wiley & Sons, Inc., Hoboken, 2006.

[2] PELED, A. and LIU, B., Digital Signal Processing: Theory, Design, and Implementation, John Wiley & Sons, Inc., New York, 1976.

[3] FERNANDES, F.T., CORREA, L.C., DE NARDIN, C., LONGO, A. and FARRET, F.A., "Improved analytical solution to obtain the MPP of PV modules", 39th Annual Conference of the IEEE Industrial Electronics Society, Vienna, vol. 1, IEEE, pp. 1674–1676, November 10–13, 2013.

[4] LARMINIE, J.E. and DICKS, A., Fuel Cell Systems Explained, John Wiley & Sons, Ltd, Chichester, 2000.

[5] CORRÊA, J.M., FARRET, F.A., CANHA, L.N. and SIMÕES, M.G., "An electrochemical-based fuel cell model suitable for electrical engineering automation approach", IEEE Transactions on Industrial Electronics Society, vol. 51, no. 5, pp. 1103–1112, October 2004.

[6] CARNIELETTO, R., PARIZZI, J.B., FARRET, F.A. and SCHITTLER, A.C., "Evaluation of the use of secondary energy for hydrogen generation", Proceedings of the 2010 IEEE International Conference on Industrial Technology, Valparaiso, Chile, IEEE, March 14–17, 2010.

[7] RAMOS, D.B., LENZ, J.M., FERRIGOLO, F.Z. and FARRET, F.A., "Proposal of a methodology using fuzzy logic control for PEMFC efficiency improving", Proceedings of the Brazilian Congress of Automation, Pato Branco, vol. 1. pp. 1282–1288, CBA, 2012.

[8] GONZATTI, F., KUHN, V.N., FERRIGOLO, F.Z., MIOTTO, M. and FARRET, F.A., "Theoretical and practical analysis of the fuel cell integration of an energy storage plant using hydrogen", Proceedings of the 2014 11th IEEE/IAS International Conference on Industry Applications INDUSCON 2014, Juiz de Fora, IEEE, p. 1, December 7–10, 2014.

[9] WANG, C. and HASHEM, M., "Dynamic models and model validation for PEM fuel cells using electrical circuits", IEEE Transactions on Energy Conversion, vol. 20, pp. 442–451, 2005.

[10] QINGSHAN, X., NIANCHUN, W., ICHIYANAGI, K. and YUKITA, K., "PEM fuel cell modeling and parameter influences of performance evaluation", Third International Conference on Electric Utility Deregulation and Restructuring and Power Technologies, DRPT 2008, Nanjuing, vol. 1, no. 1, IEEE, April 6–9, 2008.

[11] LIMA, L.P., FARRET, F.A., RAMOS, D.B., FERRIGOLO, F.Z., STANGARLIN, H.W., TRAPP, J.G. and SERDOTTE, A.B., "PSIM mathematical tools to simulate PEM fuel cells including the power converter", Proceedings of the 35th Annual Conference of IEEE Industrial Electronics, 2009, IECON '09, Porto, IEEE, November 3–5, 2009.

[12] BORGES, F.A.A., DE MELLO, L.F., MATHIAS, L.C. and ROSÁRIO, J.M., "Complete development of a battery charger system with state-of-charge analysis", European International Journal of Science and Technology, vol. 2, no. 6, ISSN: 2304-9693, http://www.cekinfo.org.uk/EIJST, July 2013.

[13] TREMBLAY, O. and DESSAINT, L.A., "Experimental validation of a battery dynamic model for EV applications", World Electric Vehicle Journal, vol. 3, pp. 13–16, 2009.

[14] JACKEY, R.A., "A simple, effective lead-acid battery modeling process for electrical system component selection", SAE Technical Paper (2007): 01-0778, SAE International, Warrendale, 2007.

[15] SATO, S. and KAWAMURA, A., "A new estimation method of state of charge using terminal voltage and internal resistance for lead acid battery", vol. 2, IEEE Xplore, Power Conversion Conference, 2002. PCC Osaka, January 2002.

[16] COX, L.P., "A transient-based approach to estimation of the electrical parameters of a lead-acid battery model", 2010 IEEE Energy Conversion Congress and Exposition (ECCE), Atlanta, IEEE, pp. 4238–4242, September 12–16, 2010.

[17] PLETT, G., "Battery management system algorithms for HEV battery state-of-charge and state-of health estimation", S. Zhang, Advanced Materials and Methods for Lithium-Ion Batteries, Research Signpost, Kerala, 2007.

[18] PLETT, G., "High-performance battery-pack power estimation using a dynamic cell model", IEEE Transactions on Vehicular Technology, vol. 53, no. 5, pp. 1586–1593, 2004.

[19] SIMÕES, M.G., BUSARELLO, T.D.C., BUBSHAIT, A.S., HARIRCHI, F., POMILIO, J.A. and BLAABJERG, F., "Interactive smart battery storage for a PV and wind hybrid energy management control based on conservative power theory", International Journal of Control, vol. 89, no. 4, pp. 850–870, Taylor & Francis, 10.1080/00207179.2015.1102971, 2015.

[20] CORREA, J.M., FARRET, F.A., SIMÕES, M.G., RAMOS, D.B. and FERRIGOLO, F.Z., "Aspects of the integration of alternative sources of energy for application in distributed generation systems", 2011 XI Brazilian Power Electronics Conference (COBEP 2011), Praiamar, vol. 1, IEEE, pp. 819–824, September 11–15, 2011.

10

POWER QUALITY ANALYSIS

10.1   INTRODUCTION

Modern power grid can have a large penetration of distributed generation (DG) at the distribution level. Such DG penetration affects the electric system. Therefore, in order to avoid causing damage or improper operation, there are some standards or recommendations to be followed for a safe connection. Most of the standardization does not allow DG to perform other tasks more than just injecting active power under a unit power factor. However, it is expected that in near future, with further aggregation of communication channels and advances in GPS and sensing technologies, plus a future widespread use of phasor measurement units (PMU), digital frequency recorders (DFR) and dynamic swing recorders (DSR) will make possible further functionalities of customer-based DG. The discussion of such technologies and their use are out of scope of this book.

One of the most important standards related to the interconnection of DG into the electrical distribution system is the IEEE 1547 (parts of which are also incorporated into UL 1741) [1]. IEEE 1547 establishes criteria and requirements that all DG must adhere to in order to connect to the electrical power system. The requirements shall meet at the point of common coupling (PCC), and they are applicable to all distributed resources technologies with aggregate capacity of 10 MVA or less. The criteria concern mainly with voltage and frequency variation limits, synchronization aspects, intentional and unintentional islanding, and response to abnormal conditions. The DG

Modeling Power Electronics and Interfacing Energy Conversion Systems, First Edition.
M. Godoy Simões and Felix A. Farret.
© 2017 John Wiley & Sons, Inc. Published 2017 by John Wiley & Sons, Inc.

can be connected to the distribution grid as either a direct current (DC) or an AC microgrid. In DC microgrids, the distribution of energy is made by DC and has positive and negative conductors. One of the main advantages of DC microgrids is the absence of frequency and phase dependences among AC generators, and for very high DC voltages, the losses in the line are less than in AC systems. The converters mostly found in DC microgrids are DC–DC and DC–AC. AC microgrids are very similar to the conventional power system. Therefore, it is possible to apply conventional power system operational concepts, such as power flux control, droop control, protection schemes, and fault detection. In this chapter, we discuss simulation tools for modeling fundamentals of power quality performance assessment. Power quality indices concern the electrical interaction between the main network and their customers. It can be decoupled in two main issues to be addressed: (i) the voltage quality concerns the way in which the supply voltage affects an equipment, and (ii) the current quality concerns the way in which the equipment current affects the system.

The quality problems can approach a very wide spectrum of causes and consequences. They can be broadly classified as transient, temporary, and steady-state according to their duration and intensity. The transient problems (IEC 61000, IEEE c62.41, IEEE 1159, and IEC816) are mostly related to switching of small magnitude load events and atmospheric causes (lightening strokes, wind, and rain). The temporary problems are mostly related to faults and connection of ferroresonant transformers and capacitor switching (IEC 61009, IEEE 1159). The steady-state problems are related to connection of power converters, harmonics, phase voltage unbalance, DC offset, and voltage flickering caused by frequency modulation and recurrent phenomena (IEC 61000, IEEE 519), like train transportation, welding devices, and manufacturing equipment. The steady-state problems have their main causes associated to high-voltage DC, static VAR compensators, thyristor-controlled series compensators, thyristor-controlled phase angle regulators, static compensators, static synchronous series compensators, and unified power flow controllers (also known as FACTS).

Most of the traditional power systems view of power quality emphasizes what is the *voltage quality*, that is, voltage dips, interruptions and voltage distortions. The voltage quality at the terminals of the generator is determined by other equipment connected to the grid and by events in the grid.

Any generator unit is affected by the voltage quality in the same way as any other equipment on that feeder. Figure 10.1 shows voltage quality issues and their impact

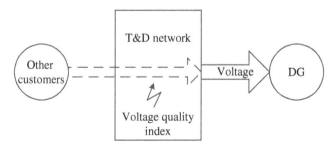

FIGURE 10.1 Voltage quality issues.

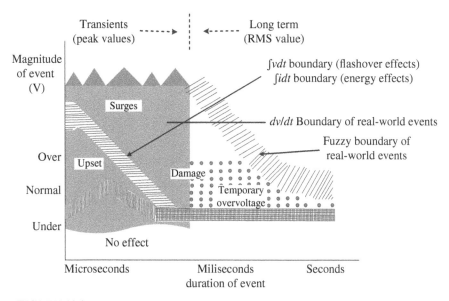

FIGURE 10.2   Voltage variations duration and corresponding effects on power systems.

caused by disturbances such as a reduction of the lifetime of equipment, communication interference, erroneous tripping, misoperation, and possible damage to equipment. An important difference between generator units and other equipment connected to the grid is that the erroneous tripping of generator units may pose a safety risk, that is, the energy flow is potentially interrupted, leading to machine overspeeding and large overvoltages. Guidelines for the immunity of generator units against voltage disturbances require a distinction between variations, normal events, and abnormal events. Figure 10.2 shows a diagram with voltage variations along their duration and corresponding effect on the power system. The study of voltage quality, voltage stability, and power quality at the transmission level is important, but not considered in this book. This is a topic pertinent to advanced power systems modeling.

The traditional view of power quality with power electronics emphasizes the *current quality*, that is, how a nonlinear load current impacts the network and how such current disturbance can affect other customers; this issue is depicted on Figure 10.3. DG can also be used to improve current quality. This requires controlling the current magnitude, the phase angle of the current compared to the voltage, and their current waveform. A good power electronic design should incorporate at the conceptual analysis solutions to impress a minimum current quality index on the power system distribution grid. It is important to understand how to measure, quantify, and evaluate power quality. That is, because DG systems when designed for distribution systems with fast controllers can improve current quality, with added active power filters (shunt), and voltage quality, with added dynamic voltage restorers (series).

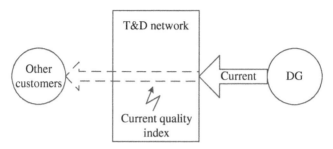

FIGURE 10.3   Current quality issues.

Power quality variations are classified as either transient or steady-state disturbances. Such perturbations pertain to abnormalities in the system voltages or currents due to faults or some abnormal operations, where steady- state variations refer to RMS deviations from their nominal quantities or harmonics. In general, disturbance analyzers, voltage recorders, and harmonic analyzers monitor such disturbances. Computer technology with advanced instrumentation can be designed for power quality monitoring and analysis. The input data for any power quality monitoring device is obtained through transducers: current transformers, voltage transformers, hall-effect current, and voltage transducers. Disturbance analyzers and disturbance monitors are instruments that are specifically designed for power quality measure-ments. There are two categories: (i) conventional analyzers and (ii) graphics-based analyzers. The conventional analyzers provide information like magnitude and dura-tion of sag/swells, under/overvoltages, while the graphic-based analyzers have memory such that real-time data can be saved to determine the source and cause of the power quality problems or used for the visualization of power grid past and real-time events.

Analyzers can also sample signals at a very high rate, such that harmonics up to about 50th order could be determined. Magnitudes of higher-order harmonics are typically much smaller than the magnitudes of lower-order harmonics. Therefore, for signal conversion and detection of higher-order harmonics, these analyzers have built-in high-resolution analog-to-digital converters. Dedicated power quality measuring instruments are manufactured to combine both functions of harmonic and disturbance measurements. The computation of power factor is given on Equation 10.1. This calcu-lation can be expanded by assuming that the voltage is purely sinusoidal and the current is distorted with a Fourier expansion. Therefore, it is necessary to understand how to calculate the harmonics of a recorded instantaneous current $i(t)$ and how to use a numerical algorithm for defining the harmonics (with their amplitude and phase) with such an implementation to calculate power factor and other indices, in order to calculate total harmonic distortion (THD) and other ones.

$$\text{PF} = \frac{P_{average}}{P_{apparent}} = \frac{\dfrac{1}{T}\displaystyle\int_0^T v(t)\cdot i(t)\,dt}{\sqrt{\dfrac{1}{T}\displaystyle\int_0^T v^2(t)\,dt}\cdot\sqrt{\dfrac{1}{T}\displaystyle\int_0^T i^2(t)\,dt}} \qquad (10.1)$$

10.2 FOURIER SERIES

Any nonsinusoidal periodic waveform $y(t)$, that is, a wave repeating with an angular frequency $\omega = 2\pi / T$ where $y(t) = y(t+T)$ can be expressed by the infinite sum indicated in Equation 10.2—one of the possible forms of a Fourier series for the summation of sines and cosines. Usually when Fourier series are solved under a pure mathematical approach, closed formulas (valid for an infinity sum) are defined. For electrical engineering applications, this formula is usually truncated, that is, solved for every harmonic, starting from the fundamental, until a given order N (say up to the 20th harmonic) because harmonics of high order are of lower amplitude and naturally filtered out by the system itself:

$$a(t) = a_0 + \sum_{n=1}^{\infty} \left[a_n \cos(n\omega t) + b_n \sin(n\omega t) \right] \tag{10.2}$$

The Fourier coefficients, a_0, a_n, and b_n, are calculated by their definitions (the reader should refer to any book on Fourier Series), as long as $y(t)$ is known in a closed mathematical form. This is very useful for some particular waveforms. Typically, those solutions are available in books, handbooks, and are available in symbolic software systems (such as Mathematica or Maple). Table 10.1 shows some typical waveforms and their corresponding Fourier series expansion, typically useful for electrical engineering applications.

After calculating the Fourier harmonic components, that is, the terms a_0, a_n, and b_n, it is necessary to visualize them. A plot of their magnitude and phase in terms of frequency is defined as *frequency response*, sometimes also called as a *Bode Diagram* (when the plot is a continuous function). The magnitude RMS and phase in radians, or degrees, per harmonic, are defined by Equations 10.3 and 10.4:

$$|A_n| = \frac{\sqrt{a_n^2 + b_n^2}}{\sqrt{2}} \tag{10.3}$$

$$\varphi_n = \tan^{-1}\left(\frac{-b_n}{a_n} \right) \tag{10.4}$$

The wave depicted in Table 10.1(a) is an idealized six-pulse waveform defined about their center of unity pulse amplitude (typical highly inductive load rectifier input current); the waveform in Table 10.1(b) is similar to (a) but displaced by an angle of $\frac{\pi}{2}$. The wave shape depicted in the Table 10.1(c) has a duration of $\frac{2\pi}{3}$, but it is displaced by an angle $\varnothing$ (it is a more general result). The waveform of Table 10.1(d) has a variable pulse width; it can range from zero (no output) to $\frac{2\pi}{3}$ for the classical bridge converter current to π for a square wave. The last entry in the table indicated by the wave shape in Table 10.1(e) is the typical line current wave into a 1 : 1 delta/wye transformer feeding a six-pulse rectifier with a DC load. The fundamental current is identical to that of a rectifier fed without a phase-shifting transformer,

TABLE 10.1 Fourier Expansion for Some Typical Power Electronic Current or Voltage Waveforms

Waveform	Fourier Series
(a)	$$f(\omega t) = \frac{2\sqrt{3}}{\pi}\left(\cos\omega t - \frac{\cos 5\omega t}{5} + \frac{\cos 7\omega t}{7} - \frac{\cos 11\omega t}{11} - \cdots\right)$$
(b)	$$f(\omega t) = \frac{2\sqrt{3}}{\pi}\left[\cos\left(\omega t - \frac{\pi}{2}\right) - \frac{\cos 5\left(\omega t - \frac{\pi}{2}\right)}{5} + \frac{\cos 7\left(\omega t - \frac{\pi}{2}\right)}{7} - \cdots\right]$$ $$= \frac{2\sqrt{3}}{\pi}\left(\sin\omega t - \frac{\sin 5\omega t}{5} - \frac{\sin 7\omega t}{7} + \frac{\sin 11\omega t}{11} - \cdots\right)$$
(c)	$$f(\omega t) = \frac{2\sqrt{3}}{\pi}\left[\sin(\omega t - \phi) - \frac{\sin 5(\omega t - \phi)}{5} - \frac{\sin 7(\omega t - \phi)}{7} + \frac{\sin 11(\omega t - \phi)}{11} - \cdots\right]$$

(d)

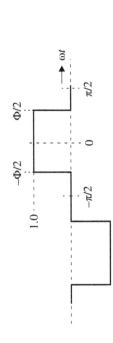

$$f(\omega t) = \sum \frac{4}{\pi n} \sin n\frac{\phi}{2}(\cos n\omega t)\quad n = 1,3,5,\ldots$$

(e)

$$f(\omega t) = \frac{2\sqrt{3}}{\pi}\left(\cos\omega t + \frac{\cos 5\omega t}{5} - \frac{\cos 7\omega t}{7} - \frac{\cos 11\omega t}{11}\ldots\right)$$

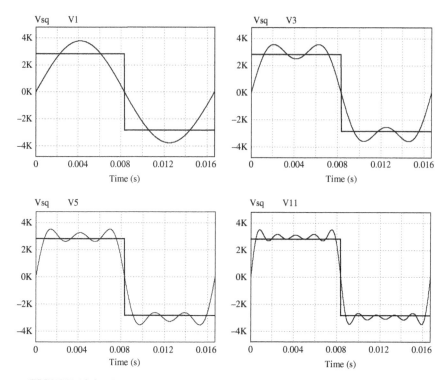

FIGURE 10.4   Quality of the square-wave representation by Fourier components.

and the shape is caused by the different phase angle of the harmonics of order $6(2k-1)\pm 1$. When developing the Fourier Series for Table 10.1(e), the coefficient

$$a_n = \frac{8}{\sqrt{3}\pi n}\left[\sin\left(n\frac{p}{3}\right)\cos\left(n\frac{p}{6}\right)\right].$$

An example of quality evolution in the wave shape analysis of a current in terms of the number of Fourier components used to represent can be observed in Figure 10.4. The square-wave harmonics represented in this figure are included up to the 11th harmonic order, taking into account that the harmonic contents of square waves have only sinusoidals of odd orders. Amplitudes of these harmonics follow mathematically the rule $V_n = V_1/n$ where the subindices refers to the harmonic orders 1st, 3rd, 5th, 7th, 9th, and 11th and all phases set at $\varphi = 0°$. As a result, when we add new components to the fundamental, the resulting wave shape gets closer and closer to the original square blocks. It is usual to represent power system waveforms up to 10 kHz since higher-order harmonics are usually filtered out by the inductances and capacitances of the transmission/distribution lines, transformers, machines, and other reactive components.

It is important to understand when to use a Fourier series versus a Fourier transform. The Fourier series is used when the time-domain variable to be expanded is periodic, and an expansion based on sines and cosines (even functions), or maybe

TABLE 10.2 **Typical Mathematical Transformations for Electrical Engineering Applications**

Transformation	Application
Fourier series	Periodic signals, oscillating systems, study of harmonics
Fourier transform	Nonrepetitive signals, transients, designing filters
Laplace transform	Full steady state and transient analysis, block diagrams, design of electronic circuits and control systems, algebraic solution of complex networks
Z transform	Discrete signals, digital signal processing

just cosines, or even based on complex exponentials is formulated. The Fourier transform is used when the time-domain variable is not periodic, that is, an infinite period. A Fourier transform converts a signal in the time domain to the frequency domain (spectrum). An inverse Fourier transform converts the frequency-domain components back into the original time-domain signal. There are a number of different mathematical transforms, which are used to analyze time functions. They are referred as frequency-domain methods. Table 10.2 shows the most common transforms and the applications that they are used. Remember that the correct representation of any waveform must contain at least two samples (Nyquist limit) within the period of the highest harmonic order to be detected by the Fourier analysis. To assure ambiguous results, it is a common practice to use 10x or higher sampling i.e. a many more sampled points than the theoretical Nyquist constraints.

For signals where both time domain and frequency domain are a sampled function (computer-based digital controllers), they can be mathematically computed as follows:

$$X(e^{j\omega}) = \sum_{n=-\infty}^{+\infty} x[n] e^{-j\omega n} \tag{10.5}$$

$$x[n] = \frac{1}{2\pi} \int_{0}^{2\pi} X(e^{j\omega}) e^{-j\omega n} d\omega \tag{10.6}$$

MATLAB has built-in functions to calculate the discrete Fourier transform (DFT) or yet fast Fourier transform (FFT) and the corresponding inverses, as depicted by Equations 10.7 and 10.8:

$$X[f_k] = \frac{1}{N} \sum_{n=0}^{N-1} x(n) e^{-\frac{j2\pi kn}{N}} \quad \text{calculated in Matlab by } \mathbf{fft}() \tag{10.7}$$

$$x(t_n) = \sum_{k=0}^{N-1} X(f_k) e^{j2\pi kn/N} \quad \text{calculated in Matlab by } \mathbf{ifft}() \tag{10.8}$$

The following MATLAB script shows an example where a square wave is recorded in a regular sampling frequency and processed by the **fft** function, with a padding of zeros to make an array with a power of 2 size (in the case in the following is 1024 points). Figure 10.5 shows the square-wave signal and its power spectrum.

(a)

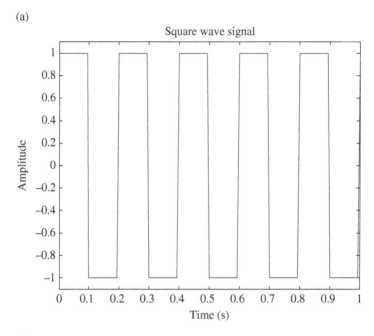

(b)

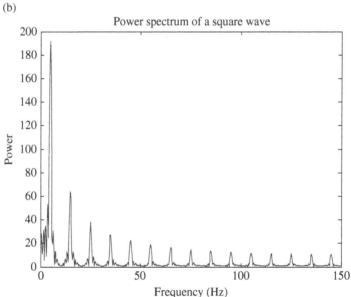

FIGURE 10.5   Waveform analysis: (a) square-wave signal and (b) their power spectrum.

The continuous function on the frequency spectrum is just a continuous plot, and the leakage from one harmonic to the next is only a mathematical effect of the sampling, called as *fence effect*, caused by a DFT length not integer of the signal frequency. The real way to interpret the **fft** solution is to observe only the magnitude and phase of each harmonic, disregarding the leakage from one harmonic to its neighbor.

```
Fs=150; %Sampling frequency
t=0:1/Fs:1; % Time Vector of 1 Second
f=5; % Create a Sine wave of f Hz
x=square(2*pi*t*f);
nfft=1024; % Length of FFT
% Take fft, padding with zeroes so that length(X) is
  equal to nfft
X=fft(x,nfft); % Take is symmetric, throw away second half
X=X(1:nfft/2); % Take the magnitude of fft of x
mx=abs(X); % Frequency vector
f=(0:nfft/2-1)*Fs/nfft; % Generate the plot, title and
  labels
figure(1);
plot(t,x);
title('Square Wave Signal');
xlabel('Time(s)');
ylabel('Amplitude');
figure(2);
plot(f,mx);
title('power Spectrum of a Square Wave');
xlabel('Frequency(Hz)');
ylabel('Power');
%%end of Matlab script -----------------------------------
```

10.3 DISCRETE FOURIER TRANSFORM FOR HARMONIC EVALUATION OF ELECTRICAL SIGNALS

The previous example discussed how the FFT function in MATLAB can be used for plotting a power spectrum of a given sampled signal. The FFT function is an implementation of a DFT, but either term can be used without any distinction in this book. This section will go through further details that must be incorporated in order to use FFT or DFT for power quality evaluation.

10.3.1   Practical Implementation Issues of DFT Using FFT

A good processing should always use the number of points N for the DFT or FFT as a power of 2 (2, 4, 8, 16, 32, 64, 128, 256, and so on). The array can be padded with zeros in order to complete a power of two. It is better to have a higher sampling time for the input data, and then discarding a few of those to make a power of 2 set for the DFT. The original waveform must be sampled at frequency, which should be much greater than the maximum frequency of the bandwidth of the signal to be evaluated. Therefore, some observations and understanding of the signal to be sampled would be valuable in determining the sampling frequency. If the original signal is too noisy, it should go through a low-pass filter in order to make sure of limited bandwidth for the **fft** analysis. If the sampling is nonuniform, maybe because there was an imperfect recording device, or asynchronous measurements, or bad data acquisition, or even a simulation with variable step size, this lack of uniform sampling must be fixed before feeding this signal for an **fft**. One solution is to preprocess the data with some interpolation techniques (or curve fitting) then construction of a fitted data with regular step size calculated from such interpolation.

The maximum frequency, or maximum bandwidth, in the spectrum analysis depends on the sampling time because of the Nyquist frequency $= f_s/2$. There is a mathematical symmetry around the sampling frequency. Therefore, it is necessary to adjust the plot for magnitude and phase of the FFT analysis within the range of $0 \leq f_m \leq f_s/2$. The DFT frequency resolution, that is, the "bin" separating each harmonic, will have a frequency band defined by f_s/N, that is, each m-harmonic is ranked in the horizontal frequency axis as shown with Equation 10.9:

$$f_{\text{harmonic analysis}}(m) = \frac{mf_s}{N} = \frac{m}{NT_s} \tag{10.9}$$

The FFT calculation performs a summation of all terms. Therefore, the values returned by FFT must be scaled. That is done with the division by the number of points, N, that is, the amplitude is corrected, depending on the total number of samplings. For a general N, the magnitude of the DFT results are directly proportional to N, and it is necessary to multiply the harmonics (not the DC value, or for $m=0$), by $2/N$. Figure 10.6 illustrates a flowchart for a complete algorithm used in folding the FFT for a harmonic analysis spectrum.

Figure 10.7 shows an example of a distorted waveform, where a DFT takes the sampled load current for one period and calculates the magnitude and phase of the frequency components. It is possible to observe the 40 time-domain samples used as the input and the frequency spectrum generated from the FFT output. The spacing between points in the output spectrum is the inverse of the length of the time sample used. In this example, the length of the sample is 20 ms. Therefore, the frequency resolution Δf is 50 Hz. Each element in the frequency plot is a harmonic since the spacing is 50 Hz. Half the number of samples used minus one gives the number of harmonics that can be resolved. Therefore, the higher the number of samples for each cycle of current, the higher the value the maximum frequency, that is, f_{max}. Since forty samples were used, the maximum harmonic that can be resolved in this case is the 19th or 950 Hz.

Frequency resolution is increased by sampling for a longer time. Having more samples during a given interval increases the maximum frequency. Removal of the fundamental from the input current is performed by setting the frequency component for 50 Hz to zero and then performing the inverse fast Fourier transform (IFFT) on each cycle of current. The IFFT recreates a time-domain signal based on the magnitude and phase information of each harmonic. The FFT must be calculated over a complete cycle to prevent spectral leakage distorting the output [2], which would lead to an incorrect compensating current signal being generated.

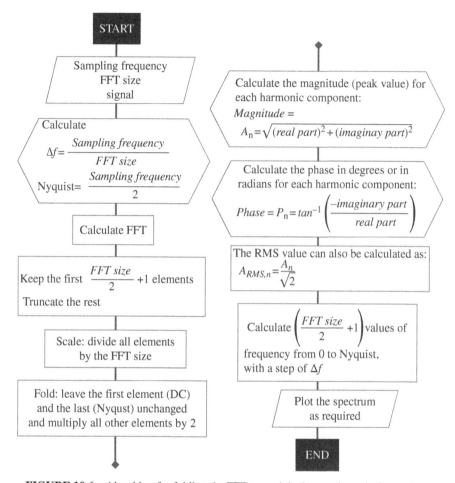

FIGURE 10.6   Algorithm for folding the FFT toward the harmonic analysis spectrum.

A frequency-domain-based harmonic isolation method has advantages over time-domain-based techniques. The magnitude of the load harmonics is known from the FFT, and this allows selective harmonic cancellation to be performed. Manipulating the harmonic magnitudes makes it possible to prevent the cancellation of certain harmonics or to reduce the compensation of individual harmonics.

10.4   ELECTRICAL POWER AND POWER FACTOR
COMPUTATION FOR DISTORTED CONDITIONS

Power factor has been defined as a ratio of mechanical or heating work (*P*) with the multiplication of effective or RMS values of voltage and current (*S*). The reason for such quotient is that the amount of steel, cooper, and dielectrics in a rotating machine

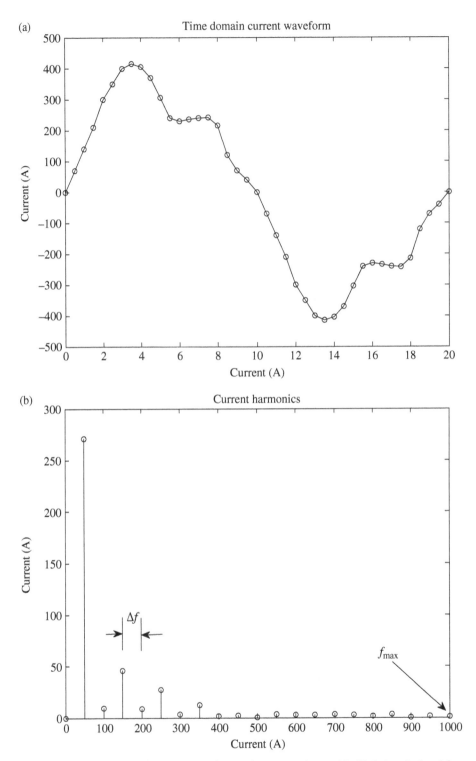

FIGURE 10.7 Time-domain samples of (a) an input waveform with (b) their calculated frequency spectrum.

or in a transformer is proportional to the apparent power. Therefore, during the beginning of the development of the utility grid, it was a very important factor to indicate costs for the generation and utility companies. In addition, losses of a power conductor carrying electrical current for energy distribution also depend on S, which is still a very important parameter, particularly for long-distance electrical power transactions.

On the other hand, the "physical work," heat or motion, executed by electricity is proportional only to the active power; in the past that was measured with analog wattmeters, and the apparent power could not be measured by such analog meters. Therefore, a definition accepted for over a century is that power factor is an index that measures the ratio of active power to the apparent power, as indicated by Equation 10.10, which can also be described as Power Factor=(Distortion Factor)×(Displacement Factor) considering low harmonic contents in the RMS voltage ($V_{RMS1} \approx V_{RMS}$):

$$PF = \frac{P}{S} = \frac{V_{s_1} I_{s_1} \cos\phi_1}{V_{RMS} I_{RMS}} = \frac{V_{RMS1}}{V_{RMS}} \cdot \frac{I_{RMS1}}{I_{RMS}} \cdot \cos\phi_1 \qquad (10.10)$$

$$PF \approx \left(\frac{I_{RMS1}}{I_{RMS}}\right)(\cos\phi_1) \qquad (10.11)$$

The power factor is strongly influenced by the harmonic content of the voltage and current. Equation 10.10 considers a case where voltage is purely sinusoidal and only the current is distorted, which is very often the approach. The harmonic distortion factors V_{RMS1}/V_{RMS} and I_{RMS1}/I_{RMS} are substituted by their THD factors and simplified for Equation 10.12. Figure 10.8 shows how average power versus apparent power will increase for higher harmonic distortion, that is, the true apparent power increases with harmonics in practical real-life power systems. The approximation (10.12) stands for the normally acceptable cases of THD of voltage and current no more than 10%. For highly distorted voltage sources, a more numerically detailed power factor computation is required. The approximation applies to acceptable cases where the THD of voltage does not exceed 10%. In addition, this definition is valid for single-phase systems or for a completely balanced three-phase system with a per-phase equivalent model. Power factor is strongly influenced by the harmonic

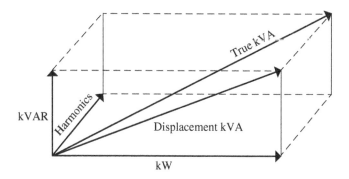

FIGURE 10.8 $P \times S \times THD$, true apparent power increases with harmonics.

content of the voltage and current. If the harmonic distortion factor I_{RMS1}/I_{RMS} is replaced by the inverse of the total current harmonic distortion factor, it can be simplified as Equation 10.12:

$$PF \approx \frac{\cos\phi_1}{\sqrt{\left(THD_{(i)}^2 + 1\right)}} \qquad (10.12)$$

If the voltage is also distorted, then a good definition of the power factor becomes mathematically more complex because it involves the product combination of each voltage order with all the other orders.

Crest factor (CF) is related to the peak rating of the power system, as indicated by Equation 10.13. The short-circuit capacity (SCC) is an important performance index for the AC source and usually associated with a maximum harmonic distortion. The AC open-circuit voltage (ideal source of voltage) multiplied by the short-circuit current of this source gives a resulting source impedance. The short-circuit ratio (SCR) is calculated by the ratio of SCC to the maximum average power, as indicated by Equation 10.14. A power system is usually considered weak when SCR < 3 and strong when SCR > 10, although such definition of weak and strong grid based on SCR is more relevant when synchronous generators are the main power sources:

$$\text{Crest factor:} \quad CF = \frac{I_{S(peak)}}{I_s} \qquad (10.13)$$

$$\text{Short-circuit ratio:} \quad SCR = \frac{SCC}{P_{ac}} \qquad (10.14)$$

Power electronics can be designed for harmonic mitigation. By having a power electronics-based interface, the power quality of the DG system improves through both control of the harmonic content of the output voltage and current. Power electronic systems can alleviate and eliminate grid reactive power requirements and at the same time restrain the power factor within a prescribed value. In addition to harmonic mitigation, power electronics can work as active filter and compensation of harmonics. A practical power quality improvement project with passive filters is suggested as laboratory project at the end of this chapter allowing further studies.

10.5   LABORATORY PROJECT: DESIGN OF A DFT-BASED ELECTRICAL POWER EVALUATION FUNCTION IN MATLAB

Write a MATLAB function (**.m**) that accepts a vector array ASCII file with consistently time sampled of a variable, plus the sampling time T_s, the phase voltage φ_V (assuming a cosine to be phase 0), and you can find the number of points N with the command size( ). The output of this script **myharmonic_analysis.m** function must be:

1. Plot (stem) of vector of amplitude of harmonic (y-axis) versus frequency (x-axis)

2. Plot (stem) of vector of phase of harmonic in degrees (*y*-axis) versus frequency (*x*-axis)

3. Plot (continuous) of power spectrum (*y*-axis) versus frequency (*x*-axis)

4. Average value

5. RMS of fundamental

6. Phase of the fundamental component, that is, $(\varphi_V - \varphi_{I_1})$

7. Total RMS

8. THD

9. Distortion factor

10. Displacement factor

11. Power factor

Test your code **myharmonic_analysis.m** with several inputs that you know, from either calculating in MATLAB, also capable of importing from PSIM, or from Matlab based Simscape Power Systems toolbox. For this design, you can test for voltage and current:

$$v(t) = \frac{480\sqrt{2}}{\sqrt{3}}\cos(\omega t) \qquad (10.15)$$

$$i(t) = \frac{2\sqrt{3}}{\pi}I_{dc}\left[\cos(\omega t) - \frac{1}{5}\cos(5\omega t) + \frac{1}{7}\cos(7\omega t) - \frac{1}{11}\cos(11\omega t) + \frac{1}{13}\cos(13\omega t)\right]$$

$$(10.16)$$

where I_{dc} is a given value of DC.

The line current in the previous equation represents the first few harmonics of a certain industrial rectifier like the one represented in Figure 10.9 whose output will look like the plot of Figure 10.10. Simulate this rectifier in PSIM and in SimPower Systems, export the data, and then compare with the results of the voltage and current mentioned previously.

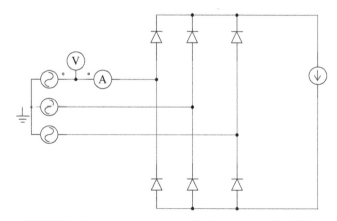

FIGURE 10.9   A three-phase rectifier implemented in PSIM.

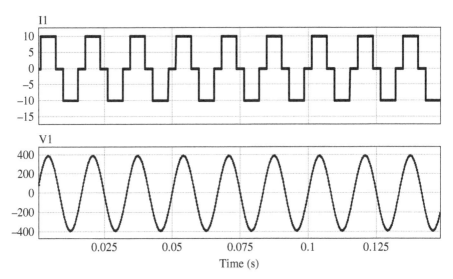

FIGURE 10.10   Current and voltage waveforms for a three-phase rectifier in PSIM.

The reader is suggested to implement rectifiers using delta–wye and delta–delta transformers, then observe the differences in harmonic content and how to improve the power factor for large industrial installations that usually use these delta–wye and delta–delta configurations.

The script function to implement the Fourier analysis is based on the flowchart in Figure 10.6. The following section shows the MATLAB code. Using the MATLAB function next, the results for a three-phase rectifier are:

```
Vrms = 277.1281
PF =   0.955
Iavg = 1.1181e-08
Irms_fundamental = 7.7980
Irms =   8.1659
DF_I =   0.9550
THD_I =   0.3108
Disp = 1

%Harmonic Analysis
clc
clear all
close all

load('rectifier.txt')
save('samples','rectifier','-ascii')
load('samples','-ascii')
t_data = samples(:,1);
```

```
V_data = samples(:,3);
I_data = samples(:,2);
Ts=t_data(2)-t_data(1);
Fs=1/Ts;
N=max(size(samples));   %sampling points
if (-1)^N==1;   %cut the second half, nfft=nfft/2
  nfft=N/2;
else
  nfft=(N-1)/2;
end
f=(0:nfft-1)*Fs/N;
%--------------------voltage------------------------%
x=V_data;
X=fft(x);
X=X(1:nfft);
mx=abs(X);
Vmx=abs(X)/(N/2);
Vmx(1)=Vmx(1)/2;
 [Vrms_fundamental I]=max(Vmx);
%Vrms_fundamental=V_fundamental/sqrt(2); %RMS of
  fundamental
f_V_fundamental=(I-1)*Fs/N; %fundamental frequency
Vangx=angle(X)/pi*180;
Vavg=Vmx(1)*cos(Vangx(1)/180*pi);  %average value
figure(1);stem(f',Vmx);title('Voltage');xlabel('Frequenc
  y(Hz)');ylabel('Magnitude');
figure(2);stem(f,Vangx);title('Voltage');xlabel('Frequen
  cy(Hz)');ylabel('Phase(degree)');
%----------------current----------------------------%
x=I_data;
X=fft(x);
X=X(1:nfft);
Imx=abs(X)/(N/2);
Imx(1)=Imx(1)/2;
 [I_fundamental I]=max(Imx);
Irms_fundamental=I_fundamental/sqrt(2); %RMS of
  fundamental
f_I_fundamental=(I-1)*Fs/N; %fundamental frequency
Iangx=angle(X)/pi*180;
Iavg=Imx(1)*cos(Iangx(1)/180*pi);  %average value
figure(3);stem(f',Imx);title('Current');xlabel('Frequency
  (Hz)');ylabel('Magnitude');
figure(4);stem(f,Iangx);title('Current');xlabel('Frequen
  cy(Hz)');ylabel('Phase(degree)');
%--------------------RMS----------------------------%
```

```
Vmx=Vmx/sqrt(2);
Vmx(1)=Vmx(1)*sqrt(2);
Vrms=sqrt(Vmx'*Vmx)   %voltage RMS
Imx=Imx/sqrt(2);
Imx(1)=Imx(1)*sqrt(2);
Irms=sqrt(Imx'*Imx);   %current RMS
%--------------------Power factor--------------------%
ph=cos((Vangx-Iangx)/180*pi);
P=Vmx'.*Imx'*ph;
S=Vrms*Irms;
PF=P/S
%------------------------------------------------------%
display(Iavg); %average value
display(Irms_fundamental); %rms of fundamental
phase_shift_fundamental=acos(ph(I));
%display(phase_shift_fundamental); %phase of fundamental
display(Irms); %total rms
DF_I=Irms_fundamental/Irms;
display(DF_I); %DF
THD_I=sqrt(1/(DF_I^2)-1);
display(THD_I);
Disp=ph(I);
display(Disp);
PF1=Disp*DF_I;
%display(PF1);
%display(PF);
%Ts
figure(5);plot(f',mx);title('Power Spectrum');xlabel('Fr
   equency(Hz)');ylabel('Power');
%%end of Matlab script --------------------------------
```

   Figure 10.11 shows the PSIM evaluation for power factor for a three-phase rectifier with harmonic measurements and of current. Figure 10.12 shows the vector of
phase of current harmonic in degrees (y-axis) versus frequency (x-axis). Figure 10.13
shows the amplitude of current harmonic (y-axis) versus frequency (x-axis),
Figure 10.14 shows the phase of voltage harmonic in degrees (y-axis) versus frequency (x-axis), and Figure 10.15 shows the plot of amplitude vector of voltage
harmonic (y-axis) versus frequency (x-axis).

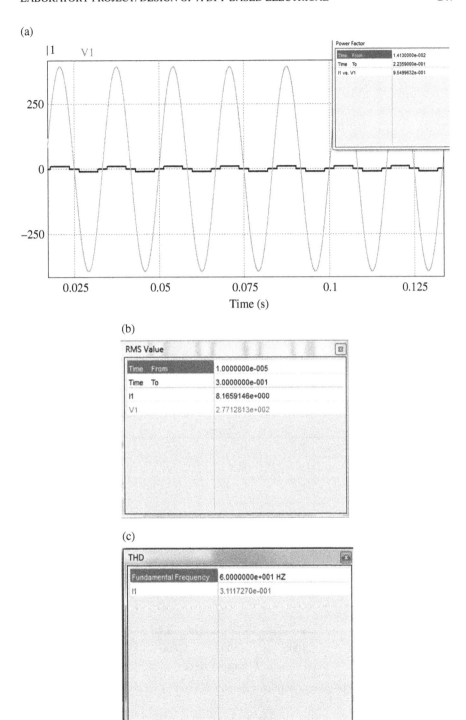

FIGURE 10.11 PSIM calculations: (a) power factor for a three-phase rectifier and waveform plot for phase, (b) harmonic measurements, and (c) THD of current.

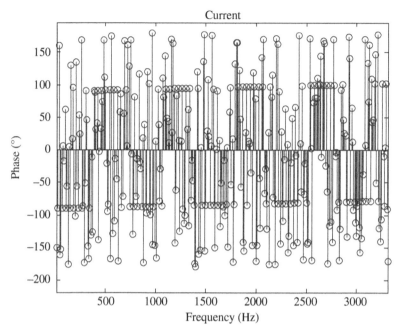

FIGURE 10.12   Plot of current harmonic phase shift (in degrees) (*y*-axis) versus frequency (*x*-axis) for the **fft** analysis.

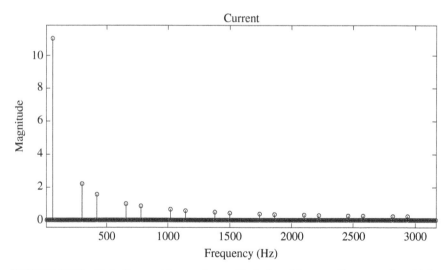

FIGURE 10.13   Plot of current harmonic amplitude (*y*-axis) versus frequency (*x*-axis) for the **fft** analysis.

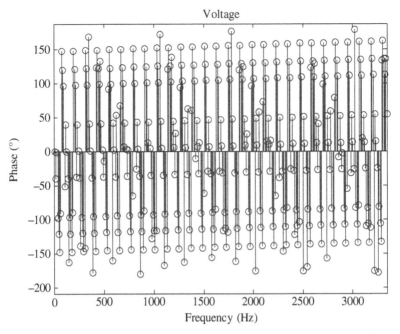

FIGURE 10.14 Plot of phase shift of voltage harmonic in degrees (*y*-axis) versus frequency (*x*-axis).

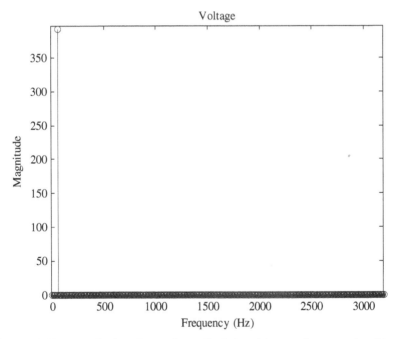

FIGURE 10.15 Plot of voltage harmonic amplitude (*y*-axis) versus frequency (*x*-axis) of the voltage amplitude vector, depicting the fundamental where all the other calculated harmonics are nearly zero in the **fft** analysis.

10.6   SUGGESTED PROBLEMS

1. The rectifier indicated in Figure 10.16 has parameters $R = 2\ \Omega$, $L = 20$ mH, $V_{DC} = 100$ V; the AC source is $v(t) = 120\sqrt{2}\cos(2\pi60t)$; and each diode has a voltage drop $V_d = 0.7$ V. Simulate in PSIM and find the power factor by measuring average power and dividing by the apparent power. Export two or three cycles of current supplied to the voltage source, with the voltage in steady-state condition and then to MATLAB; use the resulting function to generate a harmonic distortion report. Use several sampling times to export the current and show how the harmonic distortion function will behave with different sampling times. Compare PSIM versus MATLAB analysis.

2. A three-phase diode bridge rectifier with capacitive filtering at the DC link has the phase voltage and phase current waveforms as depicted in Figure 10.17. The voltage and current data for Figure 10.17 are available in a file that can be downloaded from the webpage of this book; you have to be registered with the *Google Group:* power-electronics-interfacing-energy-conversion-systems@ googlegroups.com. The simulation files and data are available after you request them to the administrator of this Google Group.

   a) Import the data in MATLAB. Use the harmonic analysis function that you developed in order to generate a harmonic distortion report.

   b) Study the three-phase industrial rectifier related to this problem and write thoughtful comments in how this circuit affects the utility distribution. Comment in how this harmonic analysis can be used to correct possible impacts of low power factor in this kind of industrial installation.

   c) Implement a three-phase rectifier with capacitive filtering that has a typical current like the previous one. Export the data from PSIM to MATLAB and use myharmonic_analysis.m to find the power factor analysis, compare with PSIM similar measurements in SimView.

   d) You have to trust your own tool, made by you, myharmonic_analysis.m. Can you use your code for any practical situation, interfacing with any software or any instrument that gives you distorted voltages and currents?

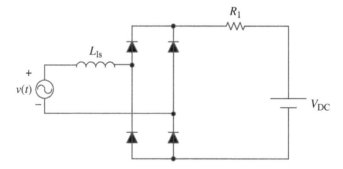

FIGURE 10.16   Single-phase rectifier with a battery connected.

(a)

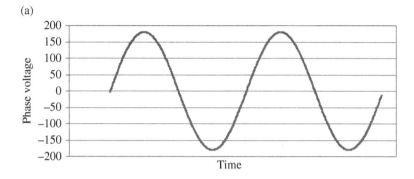

(b)

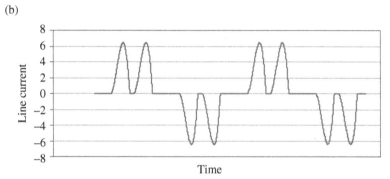

FIGURE 10.17 Three-phase rectifier with grid inductance: (a) phase voltage, (b) line current.

Describe how your tool works, and put in the beginning of your code with %, so, when you type "**help myharmonic_analysis**" in the MATLAB workspace, it explains exactly how to use such function.

3. Use the circuit of Problem #1 as an example, where you design a block diagram to be simulated in Simulink. Run Simulink for this simulation and call the function **myharmonic_analysis.m.** Therefore, you have to design a system in Simulink capable of online power factor computation using a function call to your designed harmonic analysis function. Of course, you have to run the system until the steady state settles down, in order for the **myharmonic_analysis.m** supply meaningful results.

4. A power system is represented by the Figure 10.18 in the following. There is a point of common coupling, PCC #1 of the distribution system to the transmission system at 13.8 kV, and an industrial user has a 200 kW three-phase rectifier connected to the 480V line at the PCC #2. In this project, you will have to evaluate both PCC's, showing all the indexes that quantify the power quality indexes and the corresponding power factor. You have to use a PSIM simulation, in accordance to the Figure 10.18, for this simulation-based project. Study the annexed information about passive filters, and initially design a band-pass filter trap for the 5th harmonic. Evaluate

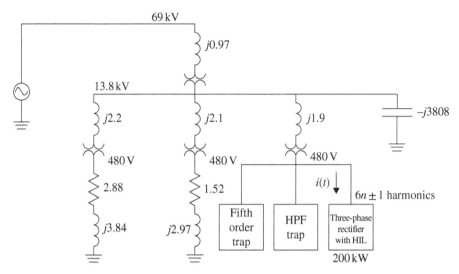

FIGURE 10.18   Power system circuit topology for power quality improvement with passive filters.

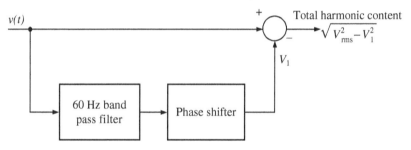

FIGURE 10.19   THD estimator using a feedforward filter.

the power quality at both PCC's with this passive filter. Then, you should design a second filter to trap the higher frequencies and evaluate the power quality with both filters. Study the IEEE Standard 519 and fine-tune your design in order to make your power system to be compliant with the harmonics in that standard. You should use the harmonics distortion analysis function that you previously designed in MATLAB in order to evaluate the power quality at both PCC. Write a report detailing the procedure and the design of the filters with their frequency response, and make a detailed analysis of how the system can be improved with the passive filters installed at the industrial user site. Write a complete report showing the methodology you used to observe the power quality and improve with passive filters selected for mitigation of harmonics.

5. Design in PSIM a feedforward filter as the one represented in Figure 10.19 to obtain the total harmonic content of a generically distorted voltage. Elaborate an example for a typical case of a power system feeding a power converter.

REFERENCES

[1] BOLLEN, M.H. and FAINAN, F., Integration of Distributed Generation in the Power System, John Wiley & Sons, Inc./IEEE Press, Hoboken/Piscataway, ISBN-10: 0470643374, ISBN-13: 978-0470643372, 2011.

[2] ARRILLAGA, J., SMITH, B.C., WATSON, N.R. and WOOD, A.R., Power System Harmonic Analysis, John Wiley & Sons, Inc., Hoboken, ISBN: 9780471975489, ISBN: 9781118878316, 2013.

FURTHER READING

BARIN, A., POZZATTI, L.F., MACHADO, R.Q., CANHA, L.N., FARRET, F.A. and ABAIDE, A.R., "Multicriteria analysis of impacts of distributed generation sources on operational network characteristics for distribution system planning concerning steady-state and transient operations", Magazine of the Brazilian Society of Power Electronics, vol. 14, pp. 75–83, 2009.

CARNIELETTO, R., BRANDÃO, D.I., SURYANARAYANAN, S., FARRET, F.A. and SIMÕES, M.G., "Smart grid initiative", IEEE Industry Applications Magazine, vol. 17, no. 5, pp. 27–35, 2011.

CHAKRABORTY, S., SIMÕES, M.G. and KRAMER, W.E., Power Electronics for Renewable and Distributed Energy Systems, 1st edition, Springer, London, 2013.

FARRET, F.A. and FRERIS, L.L., "Minimization of uncharacteristic harmonics in HVDC converters through firing angle modulation", IEE Proceedings. Generation, Transmission and Distribution, London, England, vol. 137, no. 1, pp. 45–52, 1990.

FARRET, F.A., PARIZZI, J.B., ZANCAN, M.D. and TRAPP, J.G., "Recovery of energy from harmonic filters in high power converters for simultaneous production of oxygen and hydrogen fuel", VIII Brazilian Congress of Power Electronics. Campinas, SOBRAEP. vol. 1. pp. 395–398, 2005.

IEEE Standards Coordinating Committee 21, 1547-2003, IEEE Std. IEEE Standard for Interconnecting Distributed Resources with Electric Power Systems, IEEE, New York, 2003.

REZNIK, A., SIMÕES, M.G., AL-DURRA, A. and MUYEEN, S.M., "LCL filter design and performance analysis for grid interconnected systems", IEEE Transaction on Industry Applications, vol. 50, no. 2, pp. 1225–1232, 2014.

SIMÕES, M.G. and FARRET, F.A., Modeling and Analysis with Induction Generators, 3rd edition, Taylor & Francis/CRC Press, Boca Raton, 2014.

SIMÕES, M.G., PALLE, B., CHAKRABORTY, S. and URIARTE, C., Electrical Model Development and Validation for Distributed Resources, National Renewable Energy Laboratory, Golden, 2007.

YU, X.Y., CECATI, C., DILLON, T. and SIMÕES, M.G., "The new frontier of smart grids", IEEE Industrial Electronics Magazine, vol. 5, 3, pp. 49–63, 2011.

11

FROM PSIM SIMULATION TO HARDWARE IMPLEMENTATION IN DSP

Hua Jin

11.1 INTRODUCTION

This chapter describes the basic functions and features of the simulation software PSIM. Through a grid-link inverter system example, it illustrates how to set up the power and control circuits, and how to convert analog controllers to digital controllers. Furthermore, it shows how to prepare the system for automatic code generation and processor-in-the-loop (PIL) simulation with Texas Instruments F28335 DSP.

11.2 PSIM OVERVIEW

PSIM is a simulation software specifically designed for power electronics, motor drives, and power conversion systems. It features a user-friendly interface and a very robust and fast simulation engine. In addition, its capability to generate code automatically for TI DSP and perform PIL simulation with the DSP helps to speed up the hardware implementation process significantly.

In PSIM, a circuit is represented in four parts: power circuit, control circuit, sensors, and switch controllers, as illustrated in Figure 11.1.

A power circuit includes elements such as resistors, inductors, capacitors, semiconductor switches, and some other elements that conduct electrical current. A control circuit includes blocks that handle control signals, such as multipliers,

Modeling Power Electronics and Interfacing Energy Conversion Systems, First Edition.
M. Godoy Simões and Felix A. Farret.
© 2017 John Wiley & Sons, Inc. Published 2017 by John Wiley & Sons, Inc.

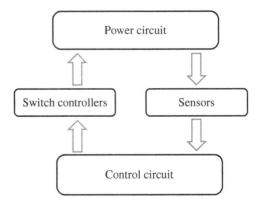

FIGURE 11.1   Circuit representation in PSIM.

dividers, s-domain, and z-domain transfer function blocks. Sensors include voltage sensors, current sensors, and speed and torque sensors. Switch controllers include blocks that control switches, such as ON–OFF controllers and PWM controllers.

The structure of the PSIM element library and key elements are listed in the following text.

Power:

- Resistors; inductors; capacitors; transformers; semiconductor switches
- Magnetic elements
- op. amp.; optocouplers; TL431
- Electric machines; encoders; speed/torque sensors; mechanical loads
- Inductors and semiconductor switches with power loss calculation capability
- Solar cells; wind turbines; batteries; ultracapacitors
- Link to finite element analysis software JMAG

Control:

- Computational function blocks
- Filters; s-domain transfer function blocks
- Logic gates
- z-domain transfer function blocks; zero-order hold; unit delay; and other function blocks for digital control
- Processor-in-the-loop block
- Links to Matlab/Simulink and ModelSim

Other:

- Voltage/current sensors
- Switch controllers
- Probes and meters

- Blocks that are common to power and control circuits, such as transformation blocks, math function blocks, DLL blocks, and C blocks
- PWM ICs (e.g., UC3842, UC3854, UCC3895); driver ICs
- Sources
- Voltage/current sources

SimCoder:

- Event control
- Library for TI F2833x floating-point DSP
- Library for TI F2803x fixed-point DSP
- TI Digital Motor Control Library

To create a circuit in PSIM, access the PSIM element library either through the Element menu or from the Library Browser. To launch the Library Browser, select **View >> Library Browser**. Note that in PSIM, one needs to define which voltages and currents to display explicitly unless the option "Save all voltage and currents during simulation" is checked. This option is under **Options >> General**, and by default, it is unchecked. To display the voltage of a node or across two nodes, use a voltage probe. To display the current of an element, either set the current flag of that element to 1 or insert a current probe.

Once the circuit schematic is completed, select **Simulate >> Simulation Control**, and place it on the schematic. Define the simulation time step and total time. Then, run the simulation by selecting **Simulate >> Run Simulation**.

After the simulation is completed, the waveform display program Simview will be launched, and waveforms can be selected for display. In Simview, one can also perform mathematical calculations (i.e., multiplication of two waveforms, and calculation of average and RMS values) and FFT analysis.

For further details on how to use PSIM, refer to *PSIM User Manual* [1].

In the following sections, a grid-link inverter example is used to illustrate how such a system can be implemented in PSIM, and how the system can be modified to generate code for TI F28335 DSP for hardware implementation.

11.3   FROM ANALOG CONTROL TO DIGITAL CONTROL

A power converter system may involve a complex control algorithm. In such a case, a DSP will be well suited to implement the control algorithm. When designing a digital controller, a common practice is to design the controller in analog domain and then convert the analog controller to the digital controller. To illustrate this process, we will use the grid-link inverter system as an example.

Figure 11.2 shows a grid-link inverter system. It consists of two stages: the DC–DC boost converter and the grid-link inverter. The boost converter provides the voltage boost and the DC bus voltage control. The inverter is connected to a load and a utility grid, and the active and reactive power flowing to the grid is regulated. Figure 11.3 shows the boost controller and the inverter controller.

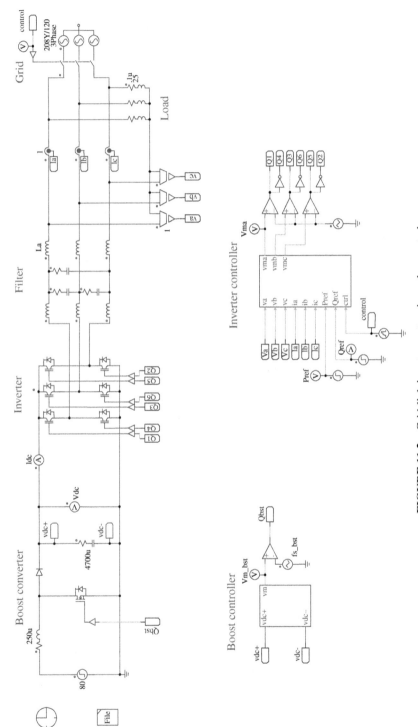

FIGURE 11.2 Grid-link inverter system in analog control.

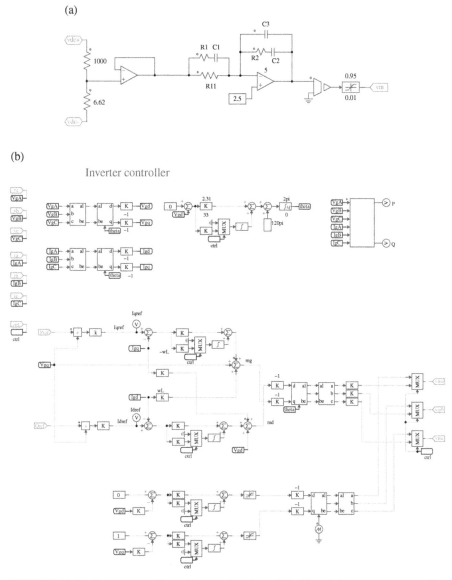

FIGURE 11.3 System controllers: (a) operational amplifier Type-3 for the boost controller and (b) analog-based three-phase inverter controller.

The boost controller uses an operational amplifier to implement a Type-3 controller to provide voltage control. The inverter controller, on the other hand, uses computational function blocks, s-domain transfer function blocks, and transformation function blocks from the PSIM's control library to implement active/reactive power control and voltage control. The inverter control algorithm is implemented in the dq-frame. A C block is used to calculate the active/reactive power.

In the inverter controller circuit, a flag "ctrl," represented by a label called "ctrl," is used to control the mode of operation. When the flag "ctrl" is set to 1, the inverter is connected to the grid and the inverter controller regulates the active/reactive power to the grid. When the flag "ctrl" is set to 0, the grid is disconnected and the load is islanded, and the inverter controller regulates the load AC voltage. A multiplexer is used to switch between these two modes of operation. Also, to avoid oversaturation of integrators, all integrator inputs are set to 0 when they are not in use. The PSIM simulation waveforms of the grid-link inverter system are shown in Figure 11.4.

The waveforms in Figure 11.4 show several transient responses. At 0.15 s, the DC bus voltage is changed from 80 to 96 V to simulate DC bus voltage change. This causes a small and short transient in the DC bus voltage but does not have notable effect on the inverter operation. At 0.2 s, the active power reference is changed from 2000 to 3000 W, and the reactive power reference is changed from 1000 to 1500 VAR. As seen from the waveforms, the inverter controller responds quickly and the active and reactive power outputs track the new references closely. At 0.3 s, the grid is disconnected and the load is islanded. In this case, the active power and reactive power are uncontrolled, and the AC load voltage is regulated instead. At 0.4 s when the grid is reconnected, the active and reactive power control resumes. In all transients, the boost controller and inverter controller are able to respond quickly, achieving good performance.

In the system in Figure 11.2, both the boost converter and the inverter are controlled by analog controllers. Since the inverter controller involves many mathematical calculations, it would be easier to implement it in a DSP. We will keep the boost controller in analog but will modify the inverter controller to a digital controller. The system with the inverter controller in digital control is shown in Figure 11.5.

As compared to the system in Figure 11.2, the system in Figure 11.5 includes additional zero-order hold blocks at the inputs of the inverter controller, and additional unit delay blocks at the outputs of the controller. In addition, inside the inverter controller, all analog integrators are replaced by z-domain digital integrators.

Designing a digital controller is an iterative process. One key factor to consider is the inherent delay. Since it takes time to perform calculation, control variables calculated in the current sampling period can be updated only at the next sampling period. This results in a delay of one sampling period, and it must be taken into account in the analog controller design. To include a delay in the analog control loop, one can use the "Time Delay" block under **Elements >> Control >> Other Function Blocks** in PSIM. After the analog controller is designed with the digital delay taken into account, the analog controller needs to be converted into a digital controller. In this example, the PI controller is implemented in such a way that the analog integrator can be directly replaced by a digital integrator. But for a general analog controller, conversion is needed. In PSIM, a tool called "s2z Converter" (under the **Utilities** menu) is provided to convert an analog controller to a digital controller. The interface of the s2z Converter is shown in Figure 11.6.

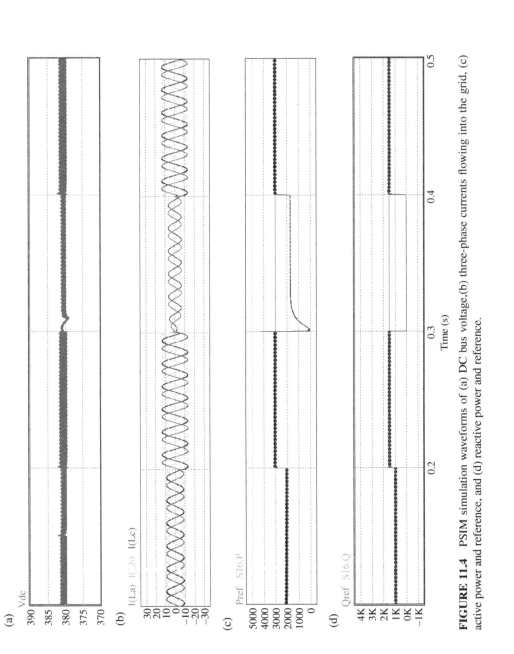

FIGURE 11.4 PSIM simulation waveforms of (a) DC bus voltage,(b) three-phase currents flowing into the grid, (c) active power and reference, and (d) reactive power and reference.

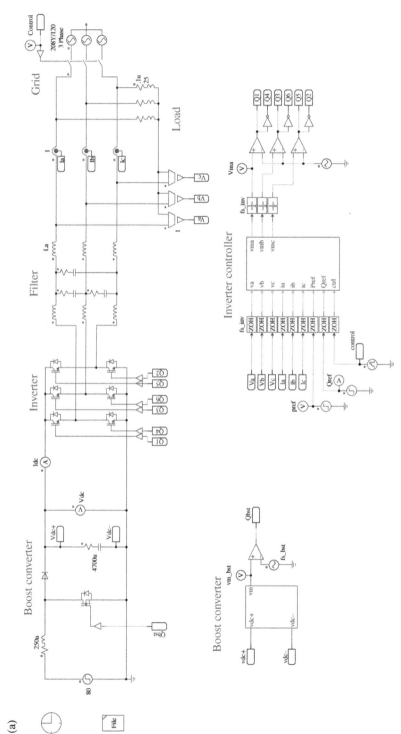

FIGURE 11.5 Grid-link inverter system with the inverter controller in digital control: (a) main system and (b) inverter digital controller.

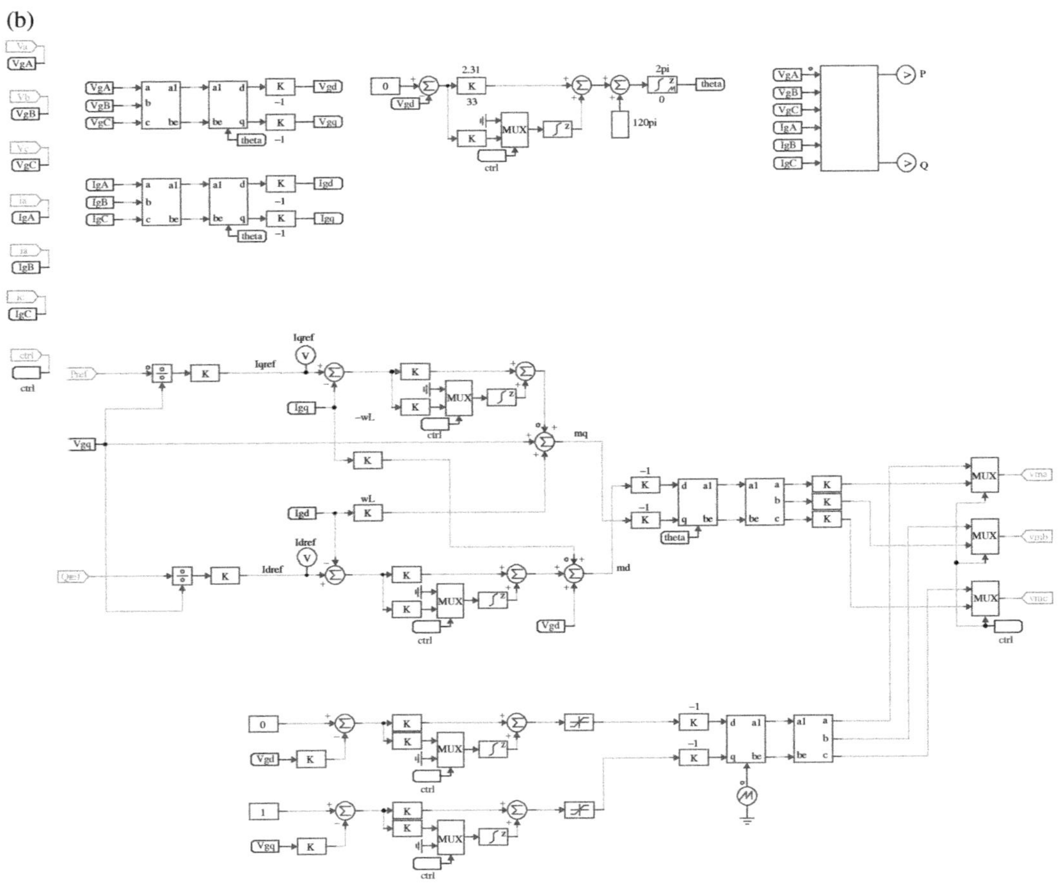

FIGURE 11.5 (Continued)

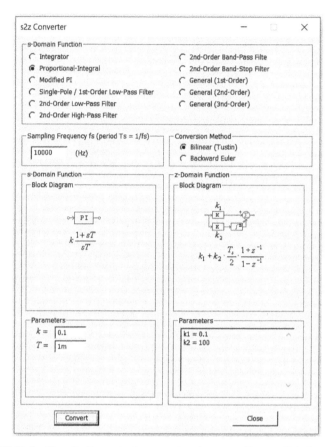

FIGURE 11.6   The s2z Converter to convert analog controllers to digital controllers.

After the sampling frequency and controller type and parameters are defined, the tool will calculate the digital controller parameters. With the inverter digital controller designed and validated, one can proceed to write the DSP code manually, or use PSIM to generate the DSP code automatically. Automatic code generation provides a very fast and reliable way of developing DSP control code. In Section 11.4, we will describe the process of using the SimCoder and hardware target library blocks in PSIM to set up the system for automatic code generation.

11.4 AUTOMATIC CODE GENERATION IN PSIM

The automatic code generation capability in PSIM allows users with little or no DSP experience to generate code and run it on DSP very quickly. The hardware target library blocks in PSIM alleviate users from the nontrivial task of setting registers and performing initialization for various DSP peripherals. For further details on how to perform automatic code generation, refer to the *SimCoder User Manual* [2] and *Tutorial: Auto Code Generation for F2833x Target* [3].

In order to prepare a system for automatic code generation for TI DSP, the following changes are needed to the schematic:

- Define the DSP type, and add the ADC converter, PWM generator, digital input/output, and other DSP peripheral blocks.
- If needed, add serial communication interface (SCI) blocks for monitoring and debugging.

One does need to have a basic understanding of the DSP hardware, peripheral block functions, and the pin layout. One should have a quick overview of the TI DSP datasheet [4], and understand the basic hardware structure. The DSP pin assignment is also described in the *SimCoder User Manual*, and simple examples are provided in PSIM for each DSP peripheral block.

11.4.1 TI F28335 DSP Peripheral Blocks

The library "F2833x Target" under **Elements >> SimCoder** contains the F28335 DSP peripheral blocks, as listed in Figure 11.7.

Key DSP peripheral blocks are the A/D converter, PWM generators, and digital input/output blocks. They will be described briefly in this section.

The F28335 DSP has 16 A/D channels, divided into Group A and Group B. The A/D converter can run in the *Continuous* mode or *Start–Stop* mode. In the Start–Stop mode, the conversion will be triggered by a PWM generator. The DSP A/D input range is from 0 to +3 V. A measured quantity can be either a DC quantity or an AC quantity. Often, a scaling circuit and an offset circuit are needed so that the signal at the A/D input is within the 0 to +3 V range. To facilitate the schematic creation, the A/D converter block in PSIM includes an offset circuit and a scaling block, and it can be defined to operate in the *DC* mode (with DC input) or *AC* mode (with AC input). To find out more on how the A/D converter is defined, refer to the relevant section of the *SimCoder User Manual*.

Figure 11.8 shows a simple test circuit for the A/D converter and the parameter dialog window.

In this circuit, a 1.5 V DC signal is read into Channel A0 in continuous mode.

The F28335 DSP provides six sets of PWM outputs (with two outputs in each set, in total 12 outputs). In addition, it provides additional 6 PWM outputs, but these outputs are shared with other peripheral blocks. The PWM function is implemented by single PWM, single phase, two-phase, and three-phase PWM generators in PSIM. In order to make it easier to set up the PWM generator, a carrier waveform similar to what is in a regular simulation is defined. Figure 11.9 shows a simple test circuit for a single-phase PWM generator and the parameter dialog window.

In this circuit, the PWM generator uses PWM 1 (general-purpose input–output GPIO0 and GPIO1) outputs. The dead time is set to 4 us, and the sampling frequency is 10 kHz. The carrier waveform is triangular with a peak value of 1. The PWM generator input is 0.6, representing a duty cycle of 0.6. Also in this circuit, the digital output pin GPIO30 is toggled at each cycle so that one can observe the start of a PWM cycle clearly.

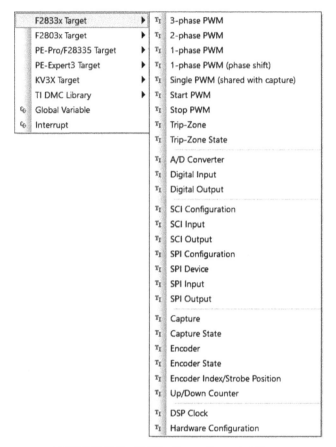

FIGURE 11.7   F2833x Target library in PSIM.

The F28335 DSP provides 88 GPIO ports that can be configured as either digital inputs or outputs. In PSIM, an eight-channel digital input block and a digital output block are provided to implement the function. A test circuit for digital input and digital output, with their parameter dialog windows, is shown in Figure 11.10.

In this circuit, two logic signals are read into Channel D0 and D1 of the digital input block. They are then sent out through the digital output block at a rate of 10 kHz. Again, the GPIO30 digital output pin is toggled at a rate of 10 kHz.

11.4.2   Adding DSP Peripheral Blocks

With the basic understanding of the DSP hardware and the PSIM hardware target library, one will be in a position to modify the grid-link inverter system for automatic code generation. First, on the SimCoder tab in Simulation Control, define the Hardware Target, project type (for Code Composer Studio), data type, and TI Digital Motor Control library version. In this example, the F2833x Target is selected, and the settings are shown in Figure 11.11.

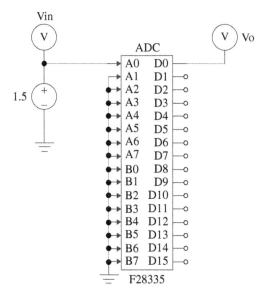

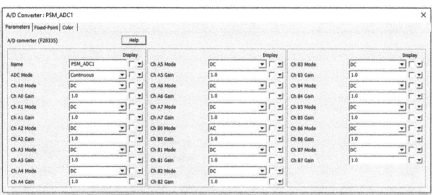

FIGURE 11.8 Test circuit for the A/D converter.

In the inverter example, an A/D converter is needed to sample three-phase AC voltages and currents, active/reactive power references, and the control signal. When defining the A/D converter, one must make sure that inputs do not exceed the limit. When a channel of the A/D converter is set to the DC mode, the input range is 0–3 V, and when it is set to the AC mode, the input range is −1.5 to 1.5 V. As shown in the system schematic in Figure 11.5, the voltage/current sensor gains are all set to 1, and the active/reactive power references are in the range of 1000–3000. They need to be scaled such that the peak values do not exceed 1.5 V for the AC mode and 3 V for the DC mode.

Based on the actual values, the current sensor gain ki is set to 0.03, and the voltage sensor gain kv is set to 0.0075. The base value of the active/reactive power references is set to 1500. The A/D converter definition is show in Figure 11.12. The A/D channel gains are set such that the variables are restored to their original values after the conversion.

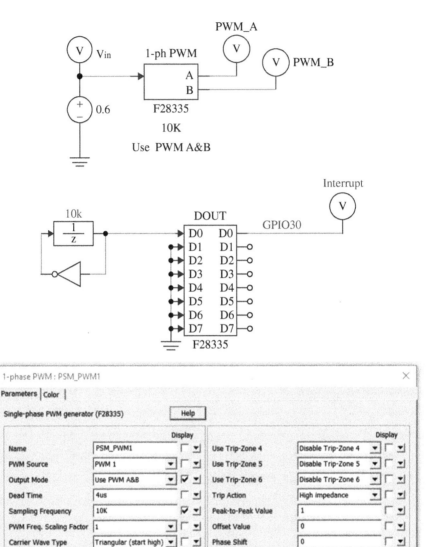

FIGURE 11.9   Test circuit for the PWM generator.

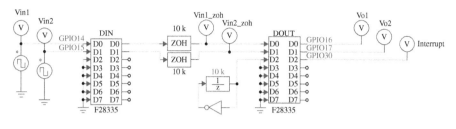

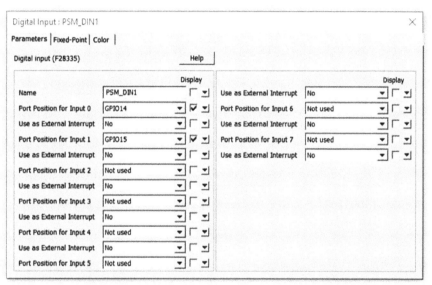

FIGURE 11.10 Test circuit for the PWM generator.

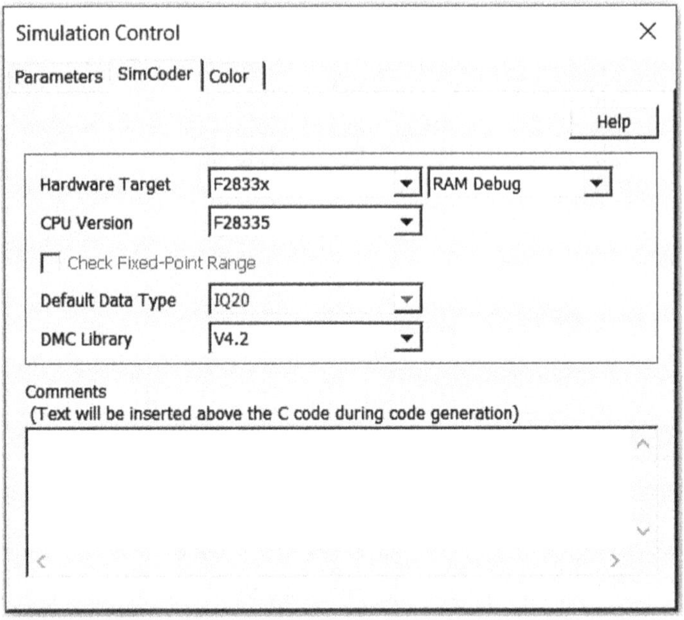

FIGURE 11.11   Setting in Simulation Control for automatic code generation.

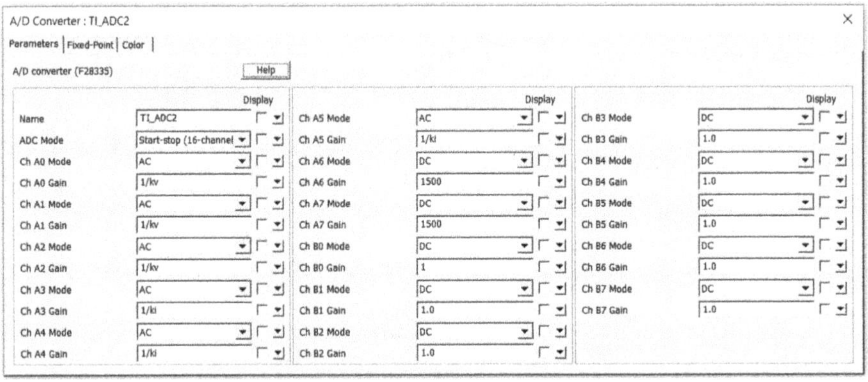

FIGURE 11.12   Definition of the A/D converter.

A PWM generator from the F2833x Target library will replace the PWM generation circuit of comparators and a carrier wave. The definition of the PWM generator
is shown in Figure 11.13.

When the parameter "Trigger ADC" is set to "Trigger ADC Group A&B", at the
beginning of each sampling cycle, the A/D converter will perform the conversion.
After the conversion is completed, the interrupt service routine will start. The way
the carrier wave is set up is the same as the carrier wave in the schematic in
Figure 11.5, with the same carrier wave type, peak-to-peak value, and offset value.

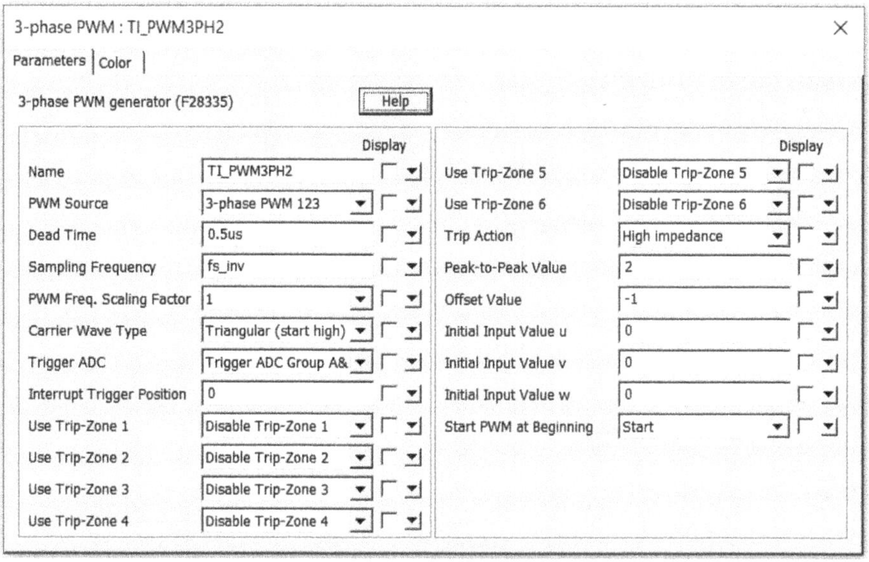

FIGURE 11.13 Definition of the PWM generator.

Besides the DSP peripheral blocks, two more changes are needed inside the controller circuit. One is the algebraic loop in the angle theta calculation. The calculation of the angle theta depends on the voltage V_{gd}, while V_{gd} itself relies on the angle theta in its calculation. This forms an algebraic loop and is not permitted in SimCoder. To break the algebraic loop, a unit delay block is inserted before the angle theta is fed back to other transformation blocks. Another change is related to dividers connected to active and reactive power references. At the very beginning, V_{gq} is 0, causing the divide-by-zero error. To avoid this problem, a C block is used to set V_{gq} to a steady-state value when it is 0.

11.4.3 Defining SCI Blocks for Real-Time Monitoring and Debugging

Debugging a digital controller is often a challenging task as variables inside DSP are not easily accessible. To solve this problem, PSIM provides a tool called "DSP Oscilloscope" (under the **Utilities** menu). Together with the SCI blocks in the Target library, the DSP Oscilloscope can monitor waveforms and change values inside DSP in real time.

To set up SCI for real-time monitoring and debugging, one needs to use the "SCI Configuration" block, "SCI Input" block, and "SCI Output" block in the F2833x Target library. The SCI Configuration block defines the SCI ports used, speed, parity check flag, and size of the buffer inside DSP to store the data. The speed and parity check settings must be the same as in the DSP Oscilloscope. For more information on how to set up SCI blocks for real-time monitoring and debugging, refer to the SCI tutorial [5].

In this example, we will monitor the angle theta and provide the option to change the reference voltage Vgq_ref and the controller gain Kp_vgq. To monitor the angle theta, connect a SCI Output block to the node of the angle theta. To change the reference

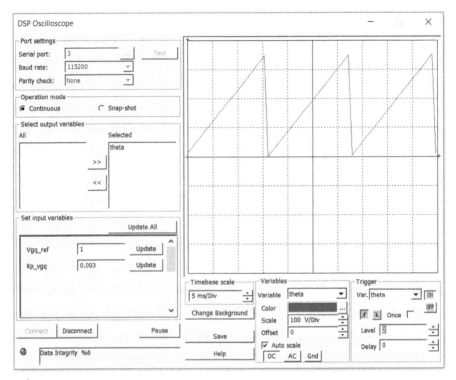

FIGURE 11.14   DSP Oscilloscope for real-time monitoring and debugging.

voltage, replace the constant with a SCI Input block with an initial value of 1. The variable Kp_vgq is defined as a global variable in the parameter file as follows:

$$(\text{Global})\,\text{Kp}\_\text{vgq} = 0.003$$

Any global variables in the DSP code will be allowed to change through the DSP Oscilloscope. Figure 11.14 shows the interface of the DSP Oscilloscope.

Once the changes to the schematic are done, to generate the code automatically, go to **Simulate>>Generate Code**. This will generate the C code that is ready to run on the F28335 DSP, and the complete project files for Code Composer Studio v3.3. If you are using a newer version of Code Composer Studio, you can use the import function to import the v3.3 project. Figure 11.15 shows the system schematic with the DSP peripheral blocks added, and Figure 11.16 shows the inverter controller with changes for SimCoder highlighted in gray.

11.5   PIL SIMULATION WITH PSIM

If one chooses to write the DSP code manually, PSIM provides a way to validate the code through PIL simulation with minimum change to the original code. With this capability, one can test and validate the code in the absence of the power converter

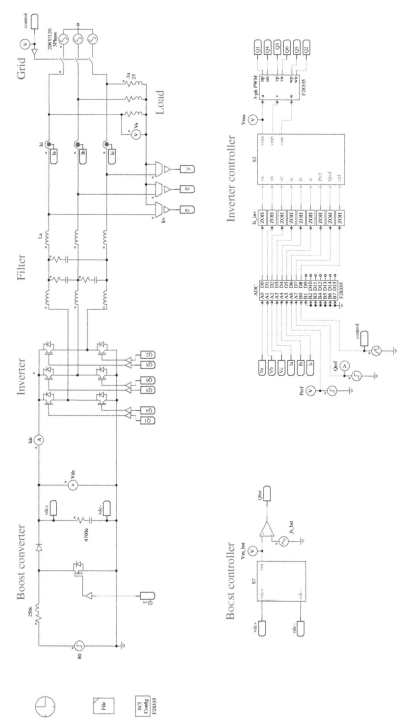

FIGURE 11.15 Grid-link inverter system with automatic code generation for F28335 DSP.

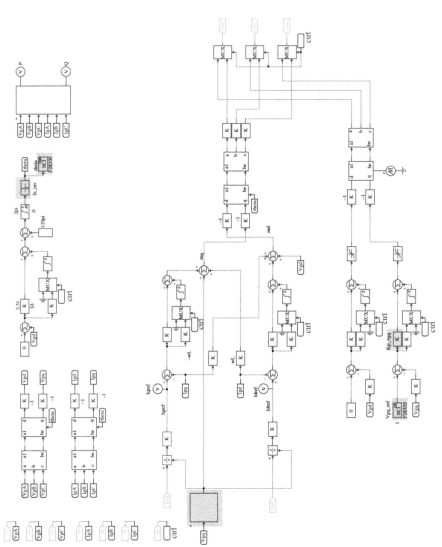

FIGURE 11.16 Inverter controller for automatic code generation.

stage. This helps to speed up the control code development, and also to perform tests and analysis that would be very difficult or impossible to do otherwise with the actual power stage. For more details on how to set up for PIL simulation, refer to the PIL tutorial [6].

To set up the PIL simulation in PSIM, go to **Elements >> Control >> PIL Module** and place a PIL block in the schematic as shown in Figure 11.17.

In order to prepare the DSP code for PIL simulation, the following changes are needed:

- Define variables that interface with PSIM as global variables.
- Comment out statements that perform A/D conversions.

The code generated from the previous section is used as the source code here. From the code, one can identify the input variables as fTI_ADC2, fTI_ADC2_1, ..., fTI_ADC2_8, and the output variables as fFunVar67, fFunVar68, and fFunVar48. These variables must be defined as global variables. Figure 11.18 shows the code before and after the changes.

The differences in the code are highlighted in gray. In the code, the function "Task()" is the interrupt service routine, and functions "PS_GetAcAdc()" and "PS_GetDcAdc()" are for A/D conversion. As shown in Figure 11.18, all input and output variables are moved out from the interrupt service routine and are defined as global variables. In addition, statements for A/D conversion are commented out. The modified code will then be compiled into the DSP executable file ".out" using Code Composer Studio. The definition of the PIL block is shown in Figure 11.19.

To launch PIL simulation, select **Simulate >> Run PSIM**. PSIM will establish the connection with the DSP hardware, upload the .out file to the DSP, and start the simulation. When simulation is running, both PSIM and DSP will run at the same time and will exchange data based on how it is defined by the PIL block.

The system in Figure 11.17 for PIL simulation is found to give the same results as the system in Figure 11.12.

11.6 CONCLUSION

This chapter provides a brief description of how to implement and simulate a system in PSIM and how to set up the system for faster hardware implementation through automatic code generation and PIL simulation. PSIM schematic files described in this chapter are available through internet download. For more information on specific details on PSIM, refer to user manuals and various PDF and online video tutorials.

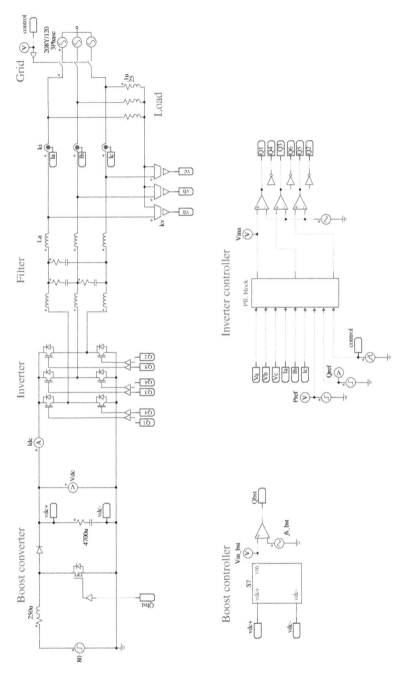

FIGURE 11.17 Grid-link inverter system with PIL simulation.

(a)

```
… … …
interrupt void Task()
{
    DefaultType          fFunVar67, fFunVar68, fFunVar48, fZOH11,
                         fTI_ADC2_8, fZOH10;
    DefaultType          fTI_ADC2_7, fZOH9, fTI_ADC2_6, fZOH8,
                         fTI_ADC2_5, fZOH7, fTI_ADC2_4;
    DefaultType          fZOH6,fTI_ADC2_3, fZOH5, fTI_ADC2_2,
                         fZOH4, fTI_ADC2_1, fZOH3;
    DefaultType          fTI_ADC2;

    PS_EnableIntr();

    fTI_ADC2 = PS_GetAcAdc(0);
    fTI_ADC2_1 = PS_GetAcAdc(1);
    fTI_ADC2_2 = PS_GetAcAdc(2);
    fTI_ADC2_3 = PS_GetAcAdc(3);
    fTI_ADC2_4 = PS_GetAcAdc(4);
    fTI_ADC2_5 = PS_GetAcAdc(5);
    fTI_ADC2_6 = PS_GetDcAdc(6);
    fTI_ADC2_7 = PS_GetDcAdc(7);
    fTI_ADC2_8 = PS_GetDcAdc(8);
    fZOH3 = fTI_ADC2;
    fZOH4 = fTI_ADC2_1;
    fZOH5 = fTI_ADC2_2;
    fZOH6 = fTI_ADC2_3;
    fZOH7 = fTI_ADC2_4;
    fZOH8 = fTI_ADC2_5;
    fZOH9 = fTI_ADC2_6;
    fZOH10 = fTI_ADC2_7;
    fZOH11 = fTI_ADC2_8;
    TaskS2(fZOH3, fZOH4, fZOH5, fZOH6, fZOH7, fZOH8, fZOH9,
fZOH10, fZOH11, &fFunVar48, &fFunVar68, &fFunVar67);

    PS_SetPwm3ph1UvwSH(fFunVar48, fFunVar68, fFunVar67);
    PS_ExitPwm1General();
}
… … …
```

(b)

```
… … …
DefaultType          fFunVar67=0, fFunVar68=0, fFunVar48=0,
                     fTI_ADC2_8=0;
DefaultType          fTI_ADC2_7=0, fTI_ADC2_6=0, fTI_ADC2_5=0,
                     fTI_ADC2_4=0;
DefaultType          fTI_ADC2_3=0, fTI_ADC2_2=0, fTI_ADC2_1=0;
DefaultType          fTI_ADC2=0;
interrupt void Task()
{
    DefaultType          fZOH11, fZOH10;
    DefaultType          fZOH9, fZOH8, fZOH7;
    DefaultType          fZOH6, fZOH5, fZOH4, fZOH3;

    PS_EnableIntr();

//    fTI_ADC2 = PS_GetAcAdc(0);
//    fTI_ADC2_1 = PS_GetAcAdc(1);
//    fTI_ADC2_2 = PS_GetAcAdc(2);
//    fTI_ADC2_3 = PS_GetAcAdc(3);
//    fTI_ADC2_4 = PS_GetAcAdc(4);
//    fTI_ADC2_5 = PS_GetAcAdc(5);
//    fTI_ADC2_6 = PS_GetDcAdc(6);
//    fTI_ADC2_7 = PS_GetDcAdc(7);
//    fTI_ADC2_8 = PS_GetDcAdc(8);
    fZOH3 = fTI_ADC2;
    fZOH4 = fTI_ADC2_1;
    fZOH5 = fTI_ADC2_2;
    fZOH6 = fTI_ADC2_3;
    fZOH7 = fTI_ADC2_4;
    fZOH8 = fTI_ADC2_5;
    fZOH9 = fTI_ADC2_6;
    fZOH10 = fTI_ADC2_7;
    fZOH11 = fTI_ADC2_8;
    TaskS2(fZOH3, fZOH4, fZOH5, fZOH6, fZOH7, fZOH8, fZOH9,
fZOH10,fZOH11, &fFunVar48, &fFunVar68, &fFunVar67);

    PS_SetPwm3ph1UvwSH(fFunVar48, fFunVar68, fFunVar67);
    PS_ExitPwm1General();
}
… … …
```

FIGURE 11.18 DSP code: (a) the original DSP code and (b) the code after the change for PIL simulation.

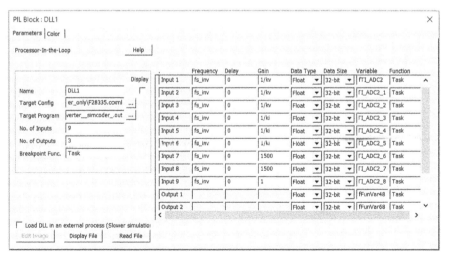

FIGURE 11.19 Definition of the PIL block.

REFERENCES

[1] *PSIM User Manual,* v10.0.4, Powersim Inc., 2015

[2] *SimCoder User Manual,* v10.0.4, Powersim Inc., 2015

[3] *Tutorial: Auto Code Generation for F2833x Target,* v10.0.4, Powersim Inc., 2015

[4] *TMS320F2833x Data Manual,* SPRS439E, Texas Instruments, 2008

[5] *Tutorial: Using SCI for Waveform Monitoring in F2833x-F2803x Target,* v10.0.4, Powersim Inc., 2015

[6] *Tutorial: Process-In-the-Loop Simulation,* v10.0.4, Powersim Inc., 2015

FURTHER READING

TI DSP Analog-to-Digital Converter (ADC) Module http://www.ti.com/lit/ug/spru812a/spru812a.pdf (accessed April 18, 2016).

TI DSP Enhanced Pulse Width Modulator (ePWM) Module http://www.ti.com/lit/ug/sprug04a/sprug04a.pdf (accessed April 18, 2016).

TI DSP System Control and Interrupts http://www.ti.com/lit/ug/sprufb0d/sprufb0d.pdf (accessed April 18, 2016).

12

DIGITAL PROCESSING TECHNIQUES APPLIED TO POWER ELECTRONICS

Danilo Iglesias Brandão and Fernando Pinhabel Marafão

12.1 INTRODUCTION

The first signal processing methods were possibly devised as analog filters and, thanks to the development of the microprocessors, the digital processing techniques have arisen. Considering the areas of power electronics and power systems, signal processing techniques have been the subject of great interest during the past decades. This is easily justified since they cover a wide field of applications, such as power quality monitoring, power quality conditioning, power electronics controlling, revenue energy metering, and power theory formulation, among many others.

Typical problems on which digital processing may support power electronics are (i) definition of power quality indices, such as total harmonic distortion (THD) or unbalance factors; (ii) selective and tuned filters used either for frequency analysis of a particular harmonic component or for selective controllers in order to provide high gain on specific frequencies, for example, resonant and repetitive controllers; (iii) accurate grid synchronization for grid-tied converters, for proper active power injection and compensation purposes; and (iv) efficient and safe operation of distributed generators, by means of maximum power point tracking (MPPT) and islanding detection techniques, among others mentioned herein.

Therefore, this chapter focuses on digital processing techniques, initially addressing basic algorithms, such as integral calculation, moving average filter (MAF), and RMS; next addressing digital filters for the identification of fundamental

Modeling Power Electronics and Interfacing Energy Conversion Systems, First Edition.
M. Godoy Simões and Felix A. Farret.
© 2017 John Wiley & Sons, Inc. Published 2017 by John Wiley & Sons, Inc.

and symmetrical components, for example, positive and negative sequences; then addressing synchronization algorithms applied to single- and three-phase systems; and finally addressing specific digital techniques for distributed generation systems, such as MPPT and islanding detection techniques. Also, some suggested problems are proposed to supplement the best understanding of this chapter.

12.2   BASIC DIGITAL PROCESSING TECHNIQUES

These basic signal processing techniques are important tools to assist the implementation of simple functions, for example, average values, collective values, and active power calculation or other more complex algorithms, such as digital filters, frequency/phase detector, islanding detection, and power electronic controllers. Sections 12.2.1, 12.2.2, 12.2.3, and 12.2.4 go through integral value calculation, MAF and RMS values. These sections present useful laboratory projects covering the most discussed functions.

12.2.1   Instantaneous and Discrete Signal Calculations

In order to introduce a fundamental notation for this chapter, let us consider a generic polyphase circuit under periodic operation, which means that any voltage or current signal repeats its waveform on a regular time basis, defined as period (T). In the following text, instantaneous quantities are denoted by lowercase symbols (x), average values by lowercase symbols with an upper bar ($\bar{x}$), RMS values by uppercase symbols (X), and vector quantities or collective values by boldface symbols ($\boldsymbol{X}$). The subscript m indicates the specific m-th phase (x_m, X_m).

The definition of an *instantaneous signal* in digital signal application is a controversial issue, because of the intrinsic delay related to measuring devices and microprocessors, that is, the discretization or sampling period (T_s). Hence, it is impossible to measure and process digital signals instantaneously, in the literal sense of the word. However, it is usually possible to measure and process them in a period of time much smaller than their fundamental period. Therefore, in order to avoid any misunderstanding, when instantaneous values are mentioned in this chapter it means $T_s \ll T$.

12.2.2   Derivative and Integral Value Calculation

The derivative and integral concepts are some of the first subjects studied in most of the engineering courses. Their digital implementation is very useful to calculate, for example, derivative function ratio, energy production or consumption during a time interval, and for the implementation of derivative- and integrative-based controllers, for example, a proportional–integral–derivative (PID) controller.

The derivative value can be calculated by the difference between two successive samples divided by the sampling period, as follows:

$$y(k) = \frac{\left[x(k) - x(k-1) \right]}{T_s}. \tag{12.1}$$

Here, y is the output variable, x is the input variable, and k is the sample counter.

The discrete domain transfer function of a differentiator can be expressed as

$$y(z) = \frac{\left[x(z) - x(z) \cdot z^{-1}\right]}{T_s} \leftrightarrow \frac{y(z)}{x(z)} = \frac{1 - z^{-1}}{T_s}, \tag{12.2}$$

remembering that z^{-1} means a unit delay.

The integral function can be digitally implemented through different methods, each method showing advantages and drawbacks. Table 12.1 shows the discrete equation and transfer function for three possible methods.

As a case study, we can integrate a sinusoidal signal, for example, at the low-voltage terminals of a step-down transformer. In this case, the trapezoidal equation of Table 12.1 was written in custom C code and implemented into C block in PSIM platform and then compared with the standard discrete integrator block of PSIM, as shown in Figure 12.1. As expected, both results are identical, as shown in Figure 12.2.

TABLE 12.1   Integrator Methods: Discrete Equations and Transfer Functions

Method	Discrete Equation	Transfer Function
Backward Euler	$y(k) = y(k-1) + T_s \cdot x(k)$	$\dfrac{y(z)}{x(z)} = \dfrac{T_s}{1 - z^{-1}}$
Forward Euler	$y(k) = y(k-1) + T_s \cdot x(k-1)$	$\dfrac{y(z)}{x(z)} = \dfrac{T_s \cdot z^{-1}}{1 - z^{-1}}$
Trapezoidal	$y(k) = y(k-1) + \dfrac{T_s}{2} \cdot \left[x(k) + x(k-1)\right]$	$\dfrac{y(z)}{x(z)} = \dfrac{T_s}{2} \dfrac{1 + z^{-1}}{1 - z^{-1}}$

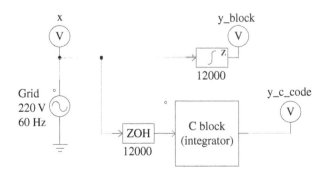

FIGURE 12.1   Trapezoidal integrator method implemented in PSIM.

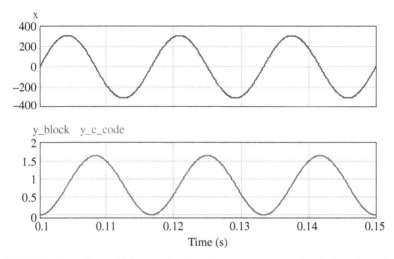

FIGURE 12.2   Sinusoidal input signal and integrated output signals (overlapped).

Customized C code implemented in C block for integrator algorithm

```
static double x;                //declaring input variable
static double y;                //declaring output variable
//declaring variables for integrator
static double intx;
static double x_prior;
static double Ts=8.33333e-5;               //sampling period
x = in[0];                                 //input signal
//integrator algorithm
intx = intx + (Ts/2)*(x + x_prior);
x_prior = x;
y = intx;
out[0] = y;                                //output signal
```

12.2.3   Moving Average Filter

The MAF can be characterized herein as quasi-instantaneous calculation, depending on its time response, which is then related to its relative vector size, for example, a MAF based on one fundamental cycle has faster dynamics than a MAF based on five cycles; however, the latter is more effective to reduce noise. At this point, we define the relative size of the MAF's vector as follows:

$$N = \frac{T}{T_s}.$$

$$(12.3)$$

(a)   (b)

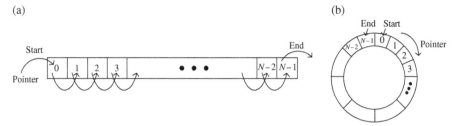

FIGURE 12.3 Linear buffer (a) and circular buffer (b) for moving average filters.

Moreover, MAF can be implemented through linear buffer, in which the start and end of the vector are not linked—as shown in Figure 12.3a, or through circular buffer where the start and end of the vector are connected, as shown in Figure 12.3b. In the former, each data is associated to a specific memory position and all the stored data must be shifted, since the pointer of the new data will always point to position 0. In the latter, after the first buffer is filled, the new data is located replacing the oldest data, so the pointer moves around a circular buffer.

The MAFs approached herein handle stiff-frequency signals, so the buffer size and sampling frequency are kept constant. Adaptive window MAF applied to variable-frequency signals can be implemented as in [1].

From Figure 12.3, either linear or circular, we note that the MAF output (y) is the sum of all data stored (S) in the buffer divided by N, as follows:

$$y(k) = \frac{S(k)}{N}, \text{where } S(k) = \sum_{j=0}^{N-1} x(k-j). \tag{12.4}$$

Thus, the discretized MAF equation can be expressed as follows:

$$y(k) = \frac{S(k-1) + x(k) - x(k-N)}{N}. \tag{12.5}$$

Here, $x(k)$ represents the current data, and $x(k-N)$ is the oldest value stored in the buffer.

From Equations 12.4 and 12.5, the discrete MAF transfer function is calculated as follows:

$$y(z) = y(z) \cdot z^{-1} + \frac{x(z) - x(z) \cdot z^{-N}}{N} \leftrightarrow \frac{y(z)}{x(z)} = \frac{1}{N} \cdot \frac{1-z^{-N}}{1-z^{-1}}. \tag{12.6}$$

Moreover, some simulation software platforms offer graphic programming, where a MAF can be discretely implemented as shown in Figure 12.4, which is based on Equation 12.6.

As a first case study, let us calculate the so-called *unbiased time integral* of the sinusoidal signal used in Section 12.2.2. By definition, the unbiased time integral ($\hat{x}$) is the integral of a periodic signal (x) without its average value [2], as follows:

$$\hat{x} = x - \bar{x} = \int_0^t x d\tau - \frac{1}{T} \cdot \int_{t-T}^t x d\tau. \tag{12.7}$$

Considering that we have already calculated the integral of a sinusoidal signal at Section 12.2.2, let us now calculate its average value ($\bar{x}$) and subtract it from the integrated signal (x_f). Figure 12.5 shows the circuit implemented in PSIM to calculate the unbiased time integral. The MAF is based on a circular buffer ($N=200$) with one fundamental cycle response, and it was implemented using two C blocks. Note that the output variable has been multiplied by the angular fundamental frequency ($2\cdot\pi\cdot60$) to normalize its magnitude. As expected, in Figure 12.6, the output variable is delayed by 90° from the input signal, and they have the same magnitude value.

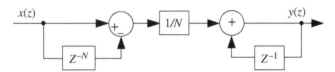

FIGURE 12.4   Discrete-time realization of a MAF.

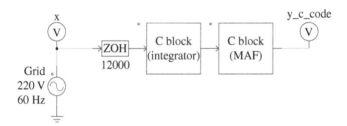

FIGURE 12.5   Unbiased time integral implemented in PSIM.

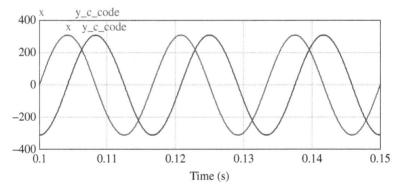

FIGURE 12.6   Sinusoidal input signal and its unbiased time integral (multiplied by w).

Customized C code implemented in C block for MAF algorithm

```
//variables for MAF
static int buffer_size = 200;          //buffer size
static double Ts = 8.33333e-5;         //sampling period
static double pi = 3.141592;           //constant pi number
static double buffer[200], sum_buffer;
static int j;                          //counter
x = in[0];                                 //input signal

//variables initialization
if(t < 0.02)
{for (j = 0; j < buffer_size; j++)
{buffer[j] = 0;}
sum_buffer = 0;
j=0;}

//MAF algorithm
if (j == buffer_size) j = 0;
sum_buffer += x - buffer[j];
buffer[j] = x;

//unbiased time integral algorithm
y = x - (sum_buffer/buffer_size);      //MAF output
j++;                                   //counter increment
out[0] = y*2*pi*60;                 //normalized output signal
```

As a second case study, let us apply the discussed stiff MAF to calculate the RMS value of the sinusoidal signal used in Section 12.2.2. Remember that the RMS value is the square root of the mean value of the squares of a signal. In Figure 12.7, the discretized equation (12.6) was written in custom C code, implemented into C block in a PSIM platform, and compared to the standard RMS block of PSIM and with the discrete time realization of Figure 12.4. Figure 12.8 shows the RMS response

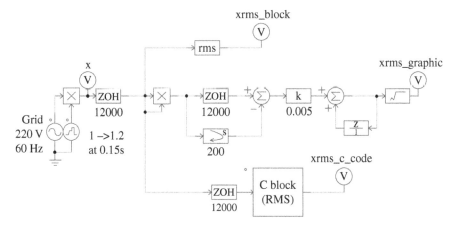

FIGURE 12.7   Comparison among the RMS implementation methods in PSIM.

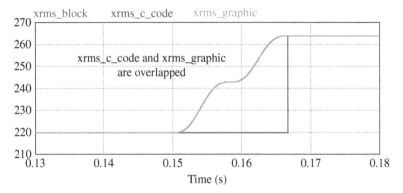

FIGURE 12.8 RMS signals.

under a step variation at 0.15 s. The standard RMS PSIM block is updated once per cycle, while the C code and graphic programming are calculated continuously and quasi-instantaneously.

Customized C code for RMS algorithm
```
//variables for RMS
static double x_rms;
x = in[0];                  //input signal
x_rms = x*x;                //squared value of input signal
//MAF algorithm
if (j == buffer_size) j = 0;
sum_buffer += x_rms - buffer[j];
buffer[j] = x_rms;
//RMS algorithm
y = sqrt(sum_buffer/buffer_size);    //MAF output
j++;                                 //counter increment
out[0] = y;                          //RMS output signal
```

12.2.4 Laboratory Project: Active Current Calculation

Based on the previous concepts, the student should implement in PSIM the electrical circuit of Figure 12.9, and should also digitally implement the C code in order to calculate the instantaneous active current, defined as

$$i_a = \frac{P}{V^2} \cdot v, \quad | \quad P = \frac{1}{T} \cdot \int_{t-T}^{t} v \cdot i \, d\tau, \tag{12.8}$$

such that v and V are the instantaneous and RMS voltage values, respectively, and P is the average active power.

As a final check, when the load has pure resistive characteristic, the active current i_a must be equal to i. If the load is set to pure inductive or pure capacitive characteristic,

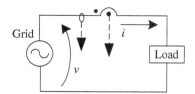

FIGURE 12.9 Electrical circuit for active current calculation.

then i_a must be null. Finally, setting an RL load, then i_a must correspond only to the power dissipated over resistance R.

12.3   FUNDAMENTAL COMPONENT IDENTIFICATION

Considering distorted voltage and current signals from nonlinear power circuits, the identification of their fundamental components is crucial for many applications, such as for the calculation of the power quality indices (THD, unbalance factors, displacement factor, etc.) or power conditioning references (active and reactive currents, synchronization signals, etc.).

Thus, the basic concept is to split a three-dimensional (three-phase) vector (x) into two main components: the sinusoidal fundamental component (x_1) and the residual component (x_{res}), which is related to harmonics, inter-harmonics, and supra-harmonics. It means

$$x = x_1 + x_{res} = \begin{bmatrix} x_{a_1} \\ x_{b_1} \\ x_{c_1} \end{bmatrix} + \begin{bmatrix} x_{ares} \\ x_{bres} \\ x_{cres} \end{bmatrix}. \tag{12.9}$$

Such decomposition is typically based on the frequency domain. However, we suggest the realization in time-domain by means of proper filtering methods.

The utilization of the word "residual" is to indicate that it represents more than the harmonic components of the original signals, and it is calculated without the need for Fast Fourier Transform (FFT). In such approach, the residual terms can be calculated by simple difference between the original signals and their fundamental or residual component, as shown in Figure 12.10.

Consequently, if one is interested in quantifying the waveform distortion of any voltage or current signal, for evaluation, compensation, or any other application, v_{res} and i_{res} should be monitored or minimized. Of course, in such case, it is not possible to identify which one and how much of the individual harmonic components are present, because it would require a complete frequency-domain analysis.

From the compensation point of view, the minimization or elimination of such residual components would mean full compensation of waveform distortions, and it would be achieved, for example, by the application of active power filters [3–5].

Regarding the implementation of the previous decomposition, in order to split the fundamental and residual components, three possible approaches are considered in

(a)   (b)

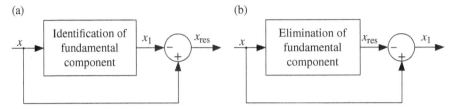

FIGURE 12.10 Concept of signal decomposition into fundamental and residual compo-nents (dual approaches). (a) Identification of fundamental component, (b) elimination of fundamental component.

Sections 12.3.1, 12.3.2, and 12.5.4, namely infinite impulse response (IIR) filters, finite impulse response (FIR) filters, and fundamental wave detector (FWD) using PLL.

12.3.1   IIR Filter

The identification or elimination of the fundamental components can be achieved, respectively, by the implementation of band-pass (Figure 12.10a) or band-stop (Figure 12.10b) digital filters, using an IIR realization scheme [6]. Assuming a band-stop filter application, the second-order IIR notch filter can be modeled in frequency domain as follows:

$$H(s) = \frac{Y(s)}{X(s)} = \frac{s^2 + \omega_0^2}{s^2 + s\omega_c + \omega_0^2} \tag{12.10}$$

Here,

$\omega_0 = 2\pi f_0$ defines the center frequency of the filter (f_0);
$\omega_c = 2\pi f_c$ defines the bandwidth of the notch filter (f_c).

The quality factor (Q) of the filter is inversely proportional to the filter's band-width and directly affects the selectivity and dynamic response of the filter. It means that the bigger the quality factor, the more selective the Bode diagram and the slower the step response of the filter.

Now, considering the application of bilinear transformation in order to achieve the discrete representation of the filter in Equation 12.10, its realization in the discrete domain results.

$$H(z) = \frac{Y(z)}{X(z)} = \frac{b_0 + b_1 z^{-1} + b_2 z^{-2}}{a_0 + a_1 z^{-1} + a_2 z^{-2}} \tag{12.11}$$

Or in terms of the difference equation,

$$a_0 y(k) = b_0 x(k) + b_1 q^{-1} x(k) + b_2 q^{-2} x(k) - a_1 q^{-1} y(k) - a_2 q^{-2} y(k)$$
$$= a_0 y(k) = b_0 x(k) + b_1 x(k-1) + b_2 x(k-2) - a_1 y(k-1) - a_2 y(k-2). \tag{12.12}$$

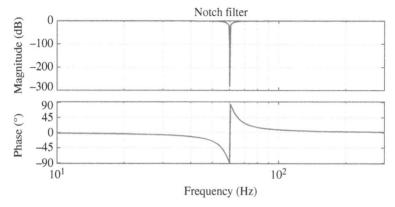

FIGURE 12.11   Frequency response of a 60 Hz notch filter.

In a practical case of a 60 Hz tuned filter, with sampling frequency of 12 kHz and stop band of 5 Hz, the frequency response is depicted as in Figure 12.11.

Thus, assuming the utilization of IIR filters for fundamental or residual components identification, some important concerns are as follows:

- Unless using adaptive IIR realizations, the tuned filters may result in considerable errors if the input signals present fundamental frequency deviations, especially in the case of narrow band filters.
- If the digital filters were implemented for limited precision discrete systems (in bits), the quantization errors may lead to unacceptable results, requiring the utilization of more sophisticated discretization techniques, such as the γ-transformation and its derivative (delta) operator [7, 8].
- The filter dynamics depends basically on its order and chosen bandwidth.

The following Matlab script shows how to design a band-pass digital filter.

```
% Analog Filter Design - 60 Hz band-pass filter
fc1 = 59;              % inferior band-pass frequency in Hz
fc2 = 61;              % superior band-pass frequency in Hz
wc1 = (2*pi*fc1);      % in rad/s
wc2 = (2*pi*fc2);
a = wc2 - wc1;         % filter bandwidth
b = wc1*wc2;
bs=[0 a 0];            % numerator H(s)
as=[1 a b];            % denominator H(s)
%     Digital Filter in Z-domain
% using shift operator by means of bilinear transformation
fp = 60;                    % pre-warping frequency to
                            the bilinear transformation
```

```
Fs = 12000;                        % 12 kHz sampling frequency
[bz,az] = bilinear(bs,as,Fs,fp)    % NUM(b) / DEN(a) of H(z)
% save original coefficients in z(shift) form
%  IIR Filter Coefficients
b0 =  0.0005232387650482284
b1 =  0.0000000000000000000
b2 = -0.0005232387650482284
a0 =  1.000000000000000
a1 = -1.997967433496970
a2 =  0.9989535224699035
```

12.3.2 FIR Filter

FIR filters are characterized by simply using current and previous input samples (not requiring prior output values such as in IIR filters). Besides, in most cases, this type of filter is considered as linear-phase filters; it means that they do not affect the phase angle of the filtered signals with respect to the original signals. A FIR filter has linear phase if and only if its coefficients are symmetric regarding to the central coefficient.

To design the FIR filter coefficients, it is possible to use a filter based on the fundamental component of the Discrete Fourier Series (DFS), which when applied to finite sequences or windows can be called Discrete Cosine Transform (DCT) [6],

$$y(k) = \frac{2}{N}\sum_{j=0}^{N-1} x(k-j)\cdot\cos\left(\frac{2\pi}{N}j\right),$$
(12.13)

where k is the sampling counter, N is the number of samples per fundamental cycle of the input signal (x), j is the circular buffer counter, and y is the filter output signal.

Equation 12.13 is implemented using the concept of moving average windows or circular buffers, resulting in a MAF, as discussed in Section 12.2.3. Consequently, the filter calculation requires a vector of dimension N, which uses the current and $N-1$ previous samples. Thus, a new sample causes the shifting of one position to all the previous vector coefficients and the exclusion of the oldest sample. Accordingly, the FIR filter can be represented in the following form:

$$y(k) = a_0 x(k) + a_1 x(k-1) + a_2 x(k-2) + \cdots + a_{N-1} x(k-(N-1)). \quad (12.14)$$

Here, x represents the sampled input signal, y the output signal, $(k-N)$ the time shifting over the samples, and a_n the filter coefficients.

It is important to note that, if needed, it is possible to design the FIR filter coefficients to work as a multiple harmonic filter. It means that the same filter may be used to identify the multiple selective harmonic frequencies of the input signal [9]. Figure 12.12 shows the frequency response of an FIR filter tuned at 60, 180, and 300 Hz.

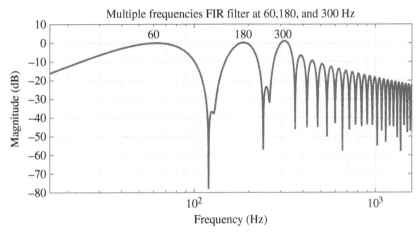

FIGURE 12.12 Frequency response of a FIR filter tuned at 60, 180, and 300 Hz.

```
% FIR FILTER DESIGN
% Fundamental frequency of 60Hz, Sampling Frequency of 12 kHz
N = 200;                  % number of samples per fundamental
                            cycle (60Hz)
SF = [1];                 % vector with the selected frequencies
                            (fundamental component)
% SF = [1 3 5];           % if required to filter the 1st,
                            3rd and 5th components
NSF = max(size(SF));      % filter dimension - number of
                            selected frequencies
coef(1:N) = zeros(1,N);   % vector initialization
for i = 1 : NSF,          % DCT calculation
    for j = 0 : N-1,
    coef(j+1) = coef(j+1) + 2/N*cos (SF(i)*(N-1-j)/N*2*pi);
    end
end
```

12.3.3   Laboratory Project: THD Calculation

As already discussed in Chapter 10, one of the most important power quality indices is the THD, which can be calculated as follows:

$$\mathrm{THD}_x = \frac{\sqrt{\sum_{h=2}^{50} X_h^2}}{X_1} = \sqrt{\sum_{h=2}^{50} \left(\frac{X_h}{X_1}\right)^2}. \qquad (12.15)$$

Such expression considers the application of a Fast Fourier Transform (FFT) up to the 50th harmonic order (h); however, in order to avoid the complex FFT calculation, a good alternative expression to the THD would be:

$$\text{THD}_x = \sqrt{\frac{1}{T} \cdot \int_{t-T}^{t} \left(\frac{x_{\text{res}}}{x_1} \right)^2 d\tau} = \frac{X_{\text{res}}}{X_1}, \tag{12.16}$$

where x_1 is the fundamental component and x_{res} represents the residual component of an input voltage or current signal, after the filtering processes discussed in Sections 12.3.1 and 12.3.2 (IIR or FIR filters).

Considering the discrete form, the THD can be achieved as follows:

$$\text{THD}_x(k) = \sqrt{\frac{1}{N} \cdot \sum_{k-N}^{k} \left[\frac{x_{\text{res}}(k)}{x_1(k)} \right]^2} = \frac{X_{\text{res}}(k)}{X_1(k)}. \tag{12.17}$$

Then, in order to implement the digital THD algorithm, the discretized equation (12.17) was implemented in PSIM by means of the customized C codes—RMS algorithm and IIR filter—and PSIM blocks (Figure 12.13). The input signal named "grid" is intentionally distorted with 10% of third harmonic and suddenly reduced to 5% at 1 s. Of course, the full THD algorithm could be implemented in one single C block from PSIM. Figure 12.14 shows the implemented THD signal response under a step variation at 1 s.

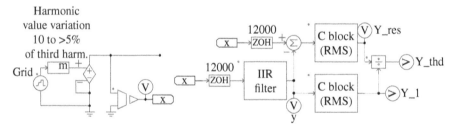

FIGURE 12.13   THD algorithm by means of IIR filter implemented in PSIM.

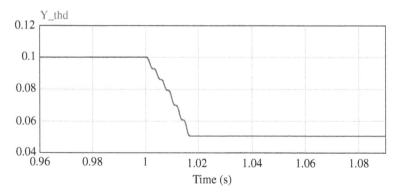

FIGURE 12.14   Implemented THD signal response.

Customized C code for IIR filter algorithm

```
static double x;                              //input
static double y;                              //output
// variables for IIR filter
static double x1, x2, y1, y2;
static double a0, a1, a2, b0, b1, b2;
x = in[0];                              //input signal - x(k)
// IIR filter algorithm
a0 = 1.000000000000000;
a1 = -1.997967433496970;
a2 = 0.9989535224699035;
b0 = 0.0005232387650482284;
b1 = 0.0000000000000000000;
b2 = -0.0005232387650482284;
y = b0*x + b1*x1 + b2*x2 - a1*y1 - a2*y2;     //eq.(12.12)
x2 = x1;              //twice delayed input signal - x(k-2)
x1 = x;               //unity delayed input signal - x(k-1)
y2 = y1;              //twice delayed output signal - y(k-2)
y1 = y;               //unity delayed output signal - y(k-1)
out[0] = y;           //output signal - y(k)
```

12.4 FORTESCUE'S SEQUENCE COMPONENTS IDENTIFICATION

In 1918, Charles Fortescue [10] demonstrated that an unbalanced three-phase system, represented in the frequency domain by three different phasors, could be decomposed into three subsystems composed of three balanced phasors: positive-, negative-, and zero-sequence symmetrical components. From that time, the symmetrical components have been used to analyze or compensate unbalanced three-phase systems. Thus, the three-phase system of positive sequence (+) would be represented as in Figure 12.15.

Considering the module of each phase component, one may observe that they are equal. It is usual to change the term $1\angle120°$ by "$\dot{a}$" or "α," known as rotational or phase displacement operator. Thus, $\dot{a} = 1\angle120°$ is understood as an operator that rotates the corresponding phasor by $120°$ in the same direction of the positive sequence.

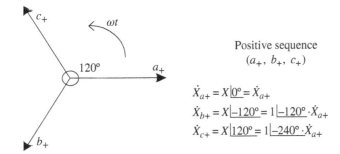

Positive sequence
(a_+, b_+, c_+)

$$\dot{X}_{a+} = X\angle 0° = \dot{X}_{a+}$$
$$\dot{X}_{b+} = X\angle -120° = 1\angle -120° \cdot \dot{X}_{a+}$$
$$\dot{X}_{c+} = X\angle 120° = 1\angle -240° \cdot \dot{X}_{a+}$$

FIGURE 12.15   Balanced system of positive sequence and corresponding phasors.

Thus, some correlated expressions are as follows:

$$\dot{X}_{a+} \quad \dot{X}_{b+} = \dot{a}^2 \cdot \dot{X}_{a+} \quad \dot{X}_{c+} = \dot{a} \cdot \dot{X}_{a+}, \tag{12.18}$$

$$\dot{a} = 1\angle 120° \quad \dot{a}^2 = 1\angle 240° = 1\angle -120° \quad 1 + \dot{a} + \dot{a}^2 = 0. \tag{12.19}$$

Now, the negative-sequence balanced three-phase system (−) would be represented as in Figure 12.16.

To make possible the algebraic manipulations with such phasors, the negative sequence would rotate in the same direction of the positive sequence, but reverse phase notation as in Figure 12.17, so that

$$\dot{X}_{a-} \quad \dot{X}_{b-} = \dot{a} \cdot \dot{X}_{a-} \quad \dot{X}_{c-} = \dot{a}^2 \cdot \dot{X}_{a-}. \tag{12.20}$$

Finally, the phasors of a zero-sequence balanced system are defined as in Equation 12.21. Figure 12.18 shows the balanced system of zero sequence components.

$$\dot{X}_{a_0} = \dot{X}_{b_0} = \dot{X}_{c_0}. \tag{12.21}$$

Therefore, some analytic expressions of the Fortescue method of symmetrical components are as follows:

$$\begin{aligned}
\dot{X}_a &= \dot{X}_{a+} + \dot{X}_{a-} + \dot{X}_{a_0}, \\
\dot{X}_b &= \dot{X}_{b+} + \dot{X}_{b-} + \dot{X}_{b_0}, \\
\dot{X}_c &= \dot{X}_{c+} + \dot{X}_{c-} + \dot{X}_{c_0}.
\end{aligned} \tag{12.22}$$

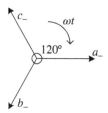

FIGURE 12.16   Balanced system of negative sequence and corresponding phasors.

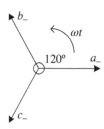

FIGURE 12.17   Balanced system of negative sequence, considering the direction of the positive sequence.

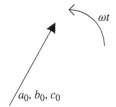

FIGURE 12.18   Balanced system of zero-sequence components.

Hence, considering that the phasors of each sequence system are balanced, it is enough to analyze a single phase, for example, phase "a" and then to expand the conclusions to the other phases. It means

$$
\begin{aligned}
\dot{X}_a &= \dot{X}_{a0} + \dot{X}_{a+} + \dot{X}_{a-}, \\
\dot{X}_b &= \dot{X}_{a0} + \dot{a}^2 \dot{X}_{a+} + \dot{a} \dot{X}_{a-}, \\
\dot{X}_c &= \dot{X}_{a0} + \dot{a} \dot{X}_{a+} + \dot{a}^2 \dot{X}_{a-}.
\end{aligned}
\tag{12.23}
$$

Assuming matrix notation we have

$$
\begin{bmatrix} \dot{X}_a \\ \dot{X}_b \\ \dot{X}_c \end{bmatrix} = \begin{bmatrix} 1 & 1 & 1 \\ 1 & \dot{a}^2 & \dot{a} \\ 1 & \dot{a} & \dot{a}^2 \end{bmatrix} \cdot \begin{bmatrix} \dot{X}_{a_0} \\ \dot{X}_{a+} \\ \dot{X}_{a-} \end{bmatrix},
\tag{12.24}
$$

so that, if one has the three sets of positive-, negative- and zero-sequence components, it is possible to achieve the original phase phasors by Equations 12.22, 12.23, and 12.24.

On the other hand, if one needs to calculate the Fortescue symmetrical components from the original three-phase variables, it is possible to do so,

$$
\begin{aligned}
\dot{X}_{a_0} &= \frac{1}{3} \cdot \left(\dot{X}_a + \dot{X}_b + \dot{X}_c \right), \\
\dot{X}_{a+} &= \frac{1}{3} \cdot \left(\dot{X}_a + \dot{a} \dot{X}_b + \dot{a}^2 \dot{X}_c \right), \\
\dot{X}_{a-} &= \frac{1}{3} \cdot \left(\dot{X}_a + \dot{a}^2 \dot{X}_b + \dot{a} \dot{X}_c \right),
\end{aligned}
\tag{12.25}
$$

or

$$
\begin{bmatrix} \dot{X}_{a_0} \\ \dot{X}_{a+} \\ \dot{X}_{a-} \end{bmatrix} = \frac{1}{3} \cdot \begin{bmatrix} 1 & 1 & 1 \\ 1 & \dot{a} & \dot{a}^2 \\ 1 & \dot{a}^2 & \dot{a} \end{bmatrix} \cdot \begin{bmatrix} \dot{X}_a \\ \dot{X}_b \\ \dot{X}_c \end{bmatrix}.
\tag{12.26}
$$

Figure 12.19 shows the three subsystems of zero, positive and negative sequence. Note that the sequence component in phases "b" and "c" can be calculated by simple operations using the shift operator $\dot{a}$.

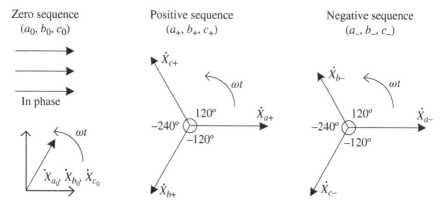

FIGURE 12.19   The three subsystems of zero, positive, and negative sequence.

Being a frequency domain method (phasors), for periodic distorted signals, the determination of the symmetrical components is possible considering the analysis of each different frequency of the original signals. However, for unbalance monitoring or compensation, it is usual to apply the Fortescue method just over the fundamental components of the voltages and currents. It means that before application of the Fortescue formulations, it is necessary to apply one of the discussed filters for the fundamental component identification.

12.4.1   Sequence Components Identification Using IIR Filter

Considering the calculation of symmetrical components over the fundamental components of measured signals, the subscript "1" is used to indicate fundamental components. Thus, supposing the calculation of sequence components over a set of unbalanced signals, for example, unbalanced three-phase voltages, and assuming the use of an IIR filter, we have to calculate the following:

$$\dot{X}_{a_1}^{+} = \frac{1}{3} \cdot \left(\dot{X}_{a_1} + \dot{X}_{b_1} e^{j\frac{2\pi}{3}} + \dot{X}_{c_1} e^{j\frac{4\pi}{3}} \right) = \frac{1}{3} \cdot \left(\dot{X}_{a_1} + \dot{a}\dot{X}_{b_1} + \dot{a}^2 \dot{X}_{c_1} \right)$$

$$= \frac{1}{3} \cdot \left(\dot{X}_{a_1} + \dot{X}_{b_1} \cdot 1\angle 120° + \dot{X}_{c_1} \cdot 1\angle 240° \right). \tag{12.27}$$

This may be attained considering that

$$X_{a_1}^{+} = \sqrt{\frac{1}{T} \cdot \int_{t-T}^{t} \left(x_{a_1}^{+} \right)^2 d\tau}, \tag{12.28}$$

and

$$x_{a_1}^{+} = \frac{1}{3} \cdot \left[x_{a_1}(t) + x_{b_1}\left(t + \frac{T}{3} \right) + x_{c_1}\left(t - \frac{T}{3} \right) \right] = \frac{1}{3} \cdot \left[x_{a_1}(t) + x_{b_1}\left(t - \frac{2T}{3} \right) + x_{c_1}\left(t - \frac{T}{3} \right) \right],$$

$$\tag{12.29}$$

or in the discrete form as follows:

$$x_{a_1}^+(k) = \frac{1}{3} \cdot \left[x_{a_1}(k) + x_{b_1}\left(k - \frac{2N}{3}\right) + x_{c_1}\left(k - \frac{N}{3}\right) \right]. \tag{12.30}$$

Thus, the negative- and zero-sequence components can be calculated, respectively, as follows:

$$x_{a_1}^-(k) = \frac{1}{3} \cdot \left[x_{a_1}(k) + x_{b_1}\left(k - \frac{N}{3}\right) + x_{c_1}\left(k - \frac{2N}{3}\right) \right], \tag{12.31}$$

$$X_{a_1}^-(k) = \sqrt{\frac{1}{N} \cdot \sum_{k-N}^{k} \left[x_{a_1}^-(k) \right]^2}, \tag{12.32}$$

$$x_{a_1}^0(k) = \frac{1}{3} \cdot \left[x_{a_1}(k) + x_{b_1}(k) + x_{c_1}(k) \right], \tag{12.33}$$

$$X_{a_1}^0(k) = \sqrt{\frac{1}{N} \cdot \sum_{k-N}^{k} \left[x_{a_1}^0(k) \right]^2}. \tag{12.34}$$

12.4.2 Sequence Component Identification Using DCT Filter

Another interesting technique used to calculate the symmetrical components is based on modifying the previously discussed DCT (FIR) filter, so that it may result in delayed output signals for direct calculation of sequence components.

So, instead of using the original filter structure as

$$
\begin{aligned}
x_{a_1}(k) &= \frac{2}{N} \cdot \sum_{j=0}^{N-1} x_a(k-j) \cdot \cos\left(\frac{2\pi}{N} j\right), \\
x_{b_1}(k) &= \frac{2}{N} \cdot \sum_{j=0}^{N-1} x_b(k-j) \cdot \cos\left(\frac{2\pi}{N} j\right), \\
x_{c_1}(k) &= \frac{2}{N} \cdot \sum_{j=0}^{N-1} x_c(k-j) \cdot \cos\left(\frac{2\pi}{N} j\right),
\end{aligned}
\tag{12.35}
$$

one could use the modified structures (') for phases "b" and "c," so that

$$
\begin{aligned}
x_{b_1}'(k) &= \frac{2}{N} \cdot \sum_{j=0}^{N-1} x_b(k-j) \cdot \cos\left[\frac{2\pi}{N}\left(j - \frac{2N}{3}\right)\right], \\
x_{c_1}'(k) &= \frac{2}{N} \cdot \sum_{j=0}^{N-1} x_c(k-j) \cdot \cos\left[\frac{2\pi}{N}\left(j - \frac{N}{3}\right)\right].
\end{aligned}
\tag{12.36}
$$

Thus, the positive-sequence can be calculated in phase "a" as follows:

$$x^+_{a_1}(k) = \frac{1}{3} \cdot \left[x_{a_1}(k) + x'_{b_1}(k) + x'_{c_1}(k) \right]. \tag{12.37}$$

Similar procedure may be applied for negative-sequence calculation. Of course, such modified filters are suitable if one is interested just in the calculation of sequence components. On the other hand, if one wants to evaluate harmonic content as well, the original FIR filter (12.13) should be used and then delayed for calculating any sequence components.

12.4.3   Laboratory Project: Calculation of Negative- and Zero-Sequence Factors

As mentioned before, the identification of positive-, negative-, and zero-sequence components is quite important to evaluate or control unbalanced three-phase systems. A common approach is to calculate the so-called negative (K^-)- and zero (K^0)-sequence factors over the voltage and current fundamental components ($h = 1$), as power quality indicators.

In this way, the negative- and zero-sequence factors can be calculated as follows:

$$K^- = \frac{X^-_{a_1}(k)}{X^+_{a_1}(k)} \quad K^0 = \frac{X^0_{a_1}(k)}{X^+_{a_1}(k)}. \tag{12.38}$$

Then, in order to implement the sequence component identification (SCI) algorithm through an IIR filter, the discretized equations (12.30, 12.31, and 12.33) were implemented in PSIM platform by means of customized C codes. SCI C block and the sequence factor were calculated using the previous customized RMS C block and PSIM blocks, as shown in Figure 12.20. Note that Equations 12.30 and 12.31 require signal shifting by one third of the buffer size. It is more effective if the

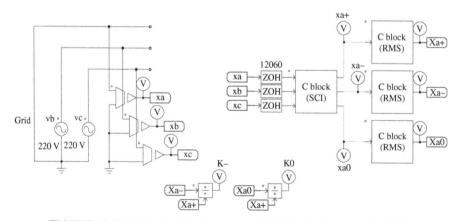

FIGURE 12.20   SCI algorithm by means of IIR filter implemented in PSIM.

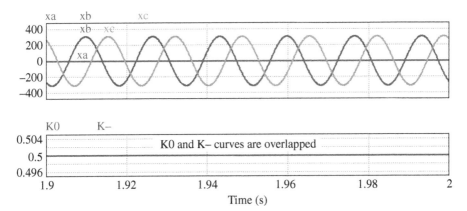

FIGURE 12.21   Unbalanced signals and corresponding unbalance factors.

buffer size is an integer number divisible by 3. Here, we have used buffer size of 201 with sampling time of 8.291874e−5 s; note that IIR filter was redesigned. The input signal named grid is asymmetrical, with the absence of phase "*a*." The results are shown in Figure 12.21, and as expected both sequence factors are equal to the simulated case.

```
Customized C code for SCI algorithm through IIR filter
static double xa, xb, xc;              //inputs
static double xap, xan, xa0;           //outputs
//....IIR filter variables
static double xa1, xa2, ya1, ya2;
static double xb1, xb2, yb1, yb2;
static double xc1, xc2, yc1, yc2;
static double a0, a1, a2, b0, b1, b2;
//....SCI filter variables
static double ya, yb, yc, yva[201], yvb[201], yvc[201];
                                       //buffer_size of 201
static int j, k1, k2;
xa = in[0];                                    //input signals
xb = in[1];
xc = in[2];
T  = 8.291874e-5 s
 s
a0 = 1;                    //IIR filter's coefficients for
a1 = −1.9979824;
a2 = 0.9989587;
b0 = 0.0005206;
b1 = 0;
b2 = −0.0005206;
k1 = j - 67;                           //delayed counter (-N/3)
k2 = j - 134;                          //delayed counter (-2N/3)
```

```
if(j-67<0) k1=j+134;
if(j-134<0) k2=j+67;
if(j >= 201) j=0;
```

```
//....IIR algorithm for three-phase
ya = b0*xa + b1*xa1 + b2*xa2 - a1*ya1 - a2*ya2; //eq.(12.12)
yb = b0*xb + b1*xb1 + b2*xb2 - a1*yb1 - a2*yb2;
yc = b0*xc + b1*xc1 + b2*xc2 - a1*yc1 - a2*yc2;
```

```
xa2 = xa1;          //twice delayed input signal - x(k-2)
xa1=xa;             //unity delayed input signal - x(k-1)
ya2 = ya1;          //twice delayed output signal - y(k-2)
ya1 = ya;           //unity delayed output signal - y(k-1)
xb2 = xb1;          //twice delayed input signal - x(k-2)
xb1=xb;             //unity delayed input signal - x(k-1)
yb2 = yb1;          //twice delayed output signal - y(k-2)
yb1 = yb;           //unity delayed output signal - y(k-1)
xc2 = xc1;          //twice delayed input signal - x(k-2)
xc1=xc;             //unity delayed input signal - x(k-1)
yc2 = yc1;          //twice delayed output signal - y(k-2)
yc1 = yc;           //unity delayed output signal - y(k-1)
yva[j] = ya;        //IIR vector output - phase a
yvb[j] = yb;        //IIR vector output - phase b
yvc[j] = yc;        //IIR vector output - phase c
```

```
//....Positive Sequence Detection
xap = (yva[j]+yvb[k2]+yvc[k1]) / 3;
```

```
//....Negative Sequence Detection
xan = (yva[j]+yvb[k1]+yvc[k2]) / 3;
```

```
//....Zero Sequence Detection
xa0 = (yva[j]+yvb[j]+yvc[j]) / 3;
```

```
out[0] = xap;                   //output sequence signals
out[1] = xan;
out[2] = xa0;
```

```
j++;                            //counter increment
```

12.5   NATURAL REFERENCE FRAME PLLS

Digital synchronization algorithms are widely used in power electronic industry and applications, as for example, in grid-connected distributed generation systems. For power system applications, the PLL algorithms are the most used, and several proposals were developed in past decades [11–13, 16]. Many of these synchronization algorithms are designed in rotational frame (dq), stationary frame ($\alpha\beta$), or in natural frame (abc), as those presented here. All of these algorithms have shown advantages and disadvantages in terms of accuracy, time response, and effectiveness with distorted voltages.

The main concept of several PLL structures is found on the orthogonality property described in Equation 12.39. Tracking the fundamental frequency (ω) and its integration to obtain the synchronization angle (θ) (12.40) ensures the orthogonality condition [16].

$$\overline{d}_p = \frac{1}{T} \cdot \int_{t-T}^{t} xx_\perp d\tau = 0, \qquad (12.39)$$

$$\theta = \int_0^t \omega \, d\tau. \qquad (12.40)$$

12.5.1 Single-Phase PLL

Based on the previous concept, this section presents a single-phase PLL structure, which is digitally implemented in Section 12.5.3. Figure 12.22 shows the single-phase PLL structure. The input periodic signal (x) is multiplied by the synthesized unitary orthogonal signal ($x_\perp$), resulting in the dot product (d_p). According to Equation 12.39, its average value should be zero in steady state. Then, the dot product error (ε_{dp}) is adjusted to zero through a PI regulator, whose output corresponds to the frequency correction ($\Delta\omega$). So, the PLL algorithm tracks the system frequency and by means of (12.40), it tracks the phase angle (θ). Thereupon, it synthesizes the unitary orthogonal signal closing the PLL control loop. Of course, the PI controller must be designed to achieve relative slow dynamics (about 5–10 Hz), since it depends on the MAF and it is usually implemented with time response of one fundamental cycle.

If the PLL should track a variable-frequency signal, as usually requested by DG systems, then a PLL structure with adaptive MAF must be implemented [1], as discussed in Section 12.2.3.

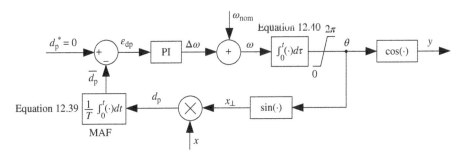

FIGURE 12.22 Block diagram of the single-phase PLL structure.

Given that the objective of this section is the discussion and application of digital processing techniques, the PI design is not deeply addressed herein. We briefly describe the simplified PLL control loop in Figure 12.23, consisting of PI controller, digital integrator, time delay (zero-order hold) and MAF. This simplification is possible because of small variations in θ yield in $\sin(\Delta\theta) \approx \Delta\theta$ [11, 12]. Moreover, the MAF can be adequately modeled as an ideal low-pass filter [13, 14]. Then, the PI controller can be designed considering the open-loop gain and phase margin of Figure 12.23, and it can be evaluated by means of traditional frequency and dynamic-response analyses [15].

Depending on the desired PLL performance and characteristics of the digital system to be used, it is demonstrated in [16] that the PLL's closed-loop transfer function may be simplified to result in the canonical form of a second-order system, in such a way that the PI parameters can be chosen in terms of the desired crossover frequency and damping factor.

12.5.2 Three-Phase PLL

The aforementioned single-phase PLL structure can be extended to a three-phase PLL structure, as shown in Figure 12.24. Note that the MAF may not be necessary, as discussed in [16], since the dot product of two orthogonal vectors may result constant and zero, if the input signals were not too much distorted and/or unbalanced, even without the MAF [16], as follows:

$$\boldsymbol{x} \cdot \boldsymbol{x}_\perp = x_a x_{a\perp} + x_b x_{b\perp} + x_c x_{c\perp} = 0. \tag{12.41}$$

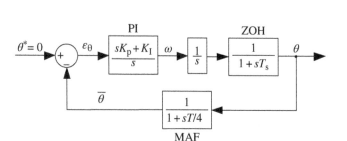

FIGURE 12.23   Block diagram of the simplified PLL control loop.

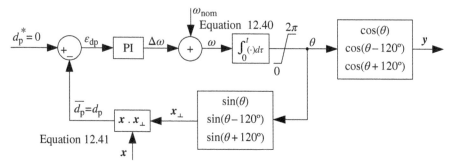

FIGURE 12.24   Block diagram of the three-phase PLL structure.

If the previous assumption is not suitable, then a general three-phase PLL structure can be set up using three independent single-phase PLLs or including the MAF after d_p calculation in Figure 12.24. Note that the three-phase plant can also be modeled as the single-phase one, with or without the MAF.

12.5.3 Laboratory Project: Single-Phase PLL Implementation

Figure 12.25 shows two possible implementation schemes of a single-phase PLL. One is based on standard analog blocks and the other on a custom C code, which is implemented into C block, both using PSIM software. To evaluate the PLL performance, the input signal named "grid" is distorted with 10% of third harmonic and a phase jump is applied at 0.3 s. Figure 12.26 shows that both implementations have practically the same results in terms of tracking the fundamental frequency, phase angle, or in phase output signal.

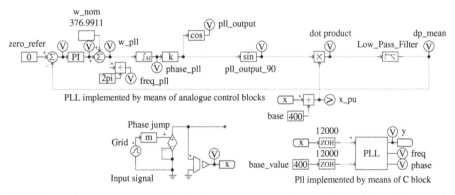

FIGURE 12.25 Single-phase PLL implemented by standard analogue PSIM blocks and using C code.

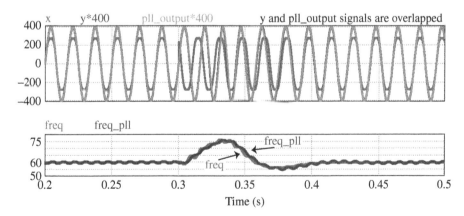

FIGURE 12.26 Performance of two different PLLs for tracking a distorted input signal and considering a phase jump at $t = 0.3$ s. Top: input signal and PLLs in phase output signals; bottom: estimated fundamental frequencies.

Customized C code for phase lock loop algorithm

```
static double a;           //input signal
static double base;        //base value - PLL algorithm works
                             with p.u. variables
static double PLL_a;       //PLL output, orthogonal to the
                             input signal
static double PLL_a1;      //PLL output, in phase with the
                             input signal
static double PLL_w;       //PLL output, frequency [rad/s]
static double PLL_phase;   //PLL output, angle [rad]

// constant variables for PLL
static double PLL_kp = 200, PLL_ki = 4000; //PIcontroller
                                             gains
static double PLL_limit = 500;    //maximum value for the
                                    proportional controller
static int PLL_ma_size = 200;     //buffer_size

// variables for PLL
static double dp, mean_out, sum_ma, Wma[200], sum_mag,
  Wmag[200], sum_rms, Wrms[200];
static int j;                     //counter for MAF

// variables for PI controller of PLL
static double PI_PLL, error;      //PI error
static double PI_kp=0, PI_ki=0;   //proportional and inte-
                                    gral part
static double PI_limit;           //integral limit
double        Tsh =8.33333e-5;    //sampling period [s]
double        pi=3.14159265359;   //constant pi value
double        w=376.99112;        //fundamental    angular
                                    frequency [rad/s]
a = in[0];                        //input signal
base = in[1];                     //base for p.u.
a = a/base;                       //input normalization

// initialization of variables
if(t<0.02)
{PLL_a = 0;
PLL_w = 0;
PLL_phase = 0;
for (j = 0; j < PLL_ma_size; j++)
{Wma[j] = 0;}
j = 0;
sum_ma = 0;
PI_ki = 0;}
```

```
// PLL algorithm
dp = a*PLL_a;        //dot product between input signal and
                         orthogonal PLL output

// MAF algorithm
if (j == PLL_ma_size) j=0;
sum_ma += dp - Wma[j];
Wma[j] = dp;
mean_out = sum_ma/PLL_ma_size;          //output mean value

// PI controller algorithm with anti-windup
error = 0 - mean_out;                          //PI error
PI_kp = error * PLL_kp;    //proportional part of the PI
                              controller
if (PI_kp > PLL_limit)     //fixed saturation - proportional
                              part
PI_kp = PLL_limit;
else if (PI_kp < -PLL_limit)
PI_kp = -PLL_limit;

if (PI_kp > 0)
PI_limit = PLL_limit - PI_kp;  //dynamic anti-windup for
                                  integral part
else
PI_limit = PLL_limit + PI_kp;
PI_ki += error * PLL_ki * Tsh; //integral  part  of  the
                                  PI controller
if (PI_ki > PI_limit)                          //anti-windup
PI_ki = PI_limit;
else if (PI_ki < -PI_limit)
PI_ki = -PI_limit;

PI_PLL = PI_kp + PI_ki;                          //PI output

PLL_w = PI_PLL + w       //nominal frequency feed-forward
PLL_phase += PLL_w*Tsh;            //Euler integrator (1/s)
if (PLL_phase >= 2*pi                 //phase angle fixed
                    saturation: 2*pi
{PLL_phase = PLL_phase - (2*pi);}
PLL_a = sin(PLL_phase);    //orthogonal signal generated
                              by the PLL
PLL_a1 = sin(PLL_phase + pi/2);    //in    phase    signal
                                      generated by the PLL
j++;                                  //increment counter
out[0] = PLL_a1;                      //in phase PLL output
out[1] = PLL_w/(2*pi);        //frequency PLL output [Hz]
out[2] = PLL_phase;                   //phase PLL output
```

12.5.4   Laboratory Project: Fundamental Wave Detector Based on PLL

In this laboratory project, the fundamental wave of a periodic signal is calculated based on the PLL structure. Thus, let us first implement the PLL algorithm in customized C code, as discussed in Section 12.5.3. If the PLL is properly programmed and designed, the in-phase PLL's output should track the input signal with a unitary magnitude. So, the fundamental waveform signal (y_1) can be calculated as follows:

$$y_1 = y \cdot c, \quad | \quad c = \frac{1}{T} \cdot \int_{t-T}^{t} 2 \cdot x \cdot y d\tau, \tag{12.42}$$

such that x is the periodic input signal, and y is the in-phase and unitary PLL output signal. Equation 12.42 can be graphically represented as in Figure 12.27.

The fundamental wave detector (FWD) was also written in custom C code and implemented into C block in PSIM platform, as shown in Figure 12.28. To evaluate the detector algorithm, the input signal named "grid" is distorted with 10% of third harmonic and a phase jump is applied at 0.3 s. Figure 12.29 shows the simulation result, on which the system is restored to steady state in 150 ms. The full implementation can be found at the webpage maintained by the authors of this book.

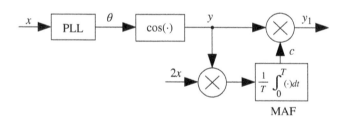

FIGURE 12.27   Block diagram of the FWD based on PLL.

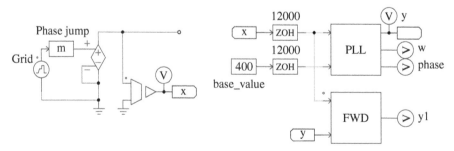

FIGURE 12.28   Single-phase PLL and FWD implemented in PSIM.

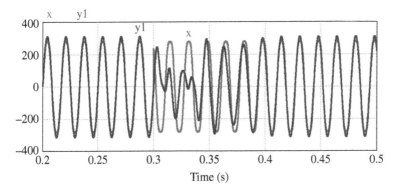

FIGURE 12.29   Input signal and in phase PLL's output (unitary signal multiplied by base value).

Customized C code implemented in C block for FWD algorithm
```
static double x, y;              //input signal
static double y1;               //output signal
```

// variables for fundamental wave detector
```
static int j, PLL_ma_size = 200;              //buffer size
static double        sum_mag, Wmag[200];
x = in[0]/400;              //normalized  input  signal  -
                             divided by base value;
y = in[1];                  //in phase unitary input signal
                             came from PLL algorithm
```

// Fundamental wave detector algorithm
```
if (j == PLL_ma_size) j = 0;
sum_mag += (2*x*y) - Wmag[j];
Wmag[j] = 2*x*y;
y1 = (sum_mag/PLL_ma_size)*y*400;
j++;

out[0] = y1;
```

12.6 MPPT TECHNIQUES

MPPT algorithms are very important in power conditioning of renewable energy resources, such as photovoltaic (PV) sources, since there is a unique point on the current–voltage (I–V) characteristic curve that corresponds to the MPP. Moreover, this point is extremely sensitive to environmental conditions that change over time; for example, the MPP of a PV solar cell varies with solar radiation and temperature. Figure 12.30 shows the I–V and power–voltage (P–V) characteristic curves of a PV solar cell, where there are three characteristic points: short-circuit point $(0, I_{sc})$, open-circuit point $(V_{oc}, 0)$, and MPP represented by (V_{MPP}, I_{MPP}). Furthermore, it is possible

(a)

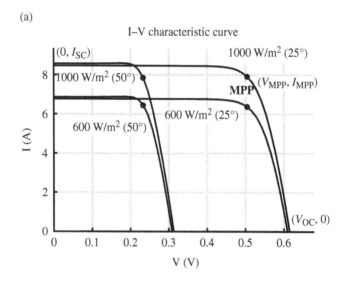

(b)

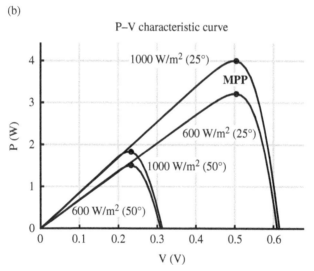

FIGURE 12.30 Voltage–current curves for solar cells: (a) I–V characteristic curve and (b) P–V characteristic curve of a typical PV solar cell.

to distinguish two operating regions: whenever the PV source operates on the left side of MPP, the PV source is more likely a constant current source; when operating on the right side of MPP, the PV source performs as constant voltage source.

PV solar cells may be associated with each other in series and parallel, in order to build up the PV solar modules. Inside a solar module, several cells can also be interconnected in series and/or parallel. The series connections are arranged to increase the voltage level, while the parallel connections are formed to increase the current capability.

The shape of the I–V and PV characteristic curves of a PV solar cell and a PV solar module is identical, except by a more accentuated relative slope along the left-hand side of the MPP. However, the occurrence of partial shading in PV modules, creating multiple power peaks on the P–V curve, are more frequent than in PV cells, because of the larger physical area of these modules. A typical occurrence of multiple peaks is shown in Figure 12.31, where local and global power peaks are observed, the latter representing the MPP. Then, the MPPT algorithms are challenged to track quickly the MPP, distinguishing between local and global peaks.

From Figure 12.31, we observe that controlling the PV output voltage (v_{PV}) or PV output current (i_{PV}) is extremely important, since it determines the PV operating point. From Figure 12.32, we note that the MPPT techniques operate together with DC–DC converters, such that the MPPT algorithm provides the reference signal,[1] that is, voltage (V_{PV}^*) or current (I_{PV}^*) reference, for the DC–DC converter's inner control

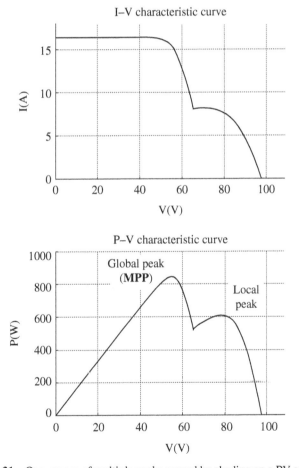

FIGURE 12.31 Occurrence of multiple peaks caused by shading on a PV solar system.

[1] The reference signal is necessary because the MPPT techniques can be applied either to voltage or current variables, depending on the converter's inner control loop design.

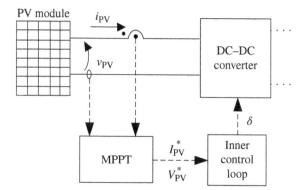

FIGURE 12.32 Typical association of MPPT algorithm with DC–DC converter in a PV power system.

loop. The inner control loop regulates the converter's input voltage (v_{pv}) or inductor current (i_{pv}) modulating the duty cycle (δ), which allows an association between the MPPT and DC–DC converter to determine the PV operating point.

Several MPPT techniques have been proposed in the literature [17], and in this section, three of the most common are presented: perturb and observe (P&O), incremental conductance (IC), and beta technique.

12.6.1 Perturb and Observe

The P&O technique, also known as hill-climbing control (HCC), periodically increments the output reference signal, increasing or decreasing the controllable PV output quantity based on the PV output power. Figure 12.33 shows the P&O flowchart applied to current, where the PV output voltage and current are measured and the active power is calculated [$\overline{p}(k)$], as described in Section 12.2.4, and compared with the previous sample $\overline{p}(k-1)$. Depending on the power and voltage comparison, the PV current reference (I_{pv}^{*}) is incremented or decremented, changing the PV operating point.

12.6.2 Incremental Conductance

The IC technique monitors the derivative of the P–V characteristic curve, on which the MPP corresponds to a derivative value equal to zero. Then, based on the signal of the derivative value, positive or negative, it is possible to recognize whether the output reference signal must be incremented or decremented, as shown in Figure 12.34.

We underline that the incremental techniques, such as P&O and IC, can be misled by the multiple power peaks of PV modules and they hardly operate exactly at the MPP; they tend to oscillate around it. Moreover, the increment must be properly chosen to provide a good commitment between the time response and the steady-state oscillation. Both techniques may be enhanced using variable increments [18].

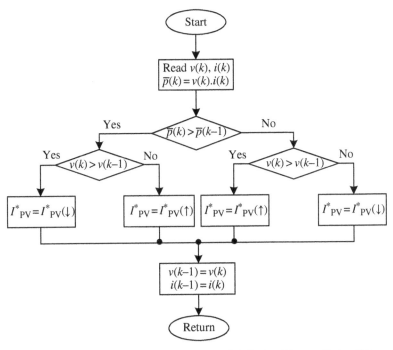

FIGURE 12.33   Flowchart of perturb and observe (P&O) algorithm applied to PV current.

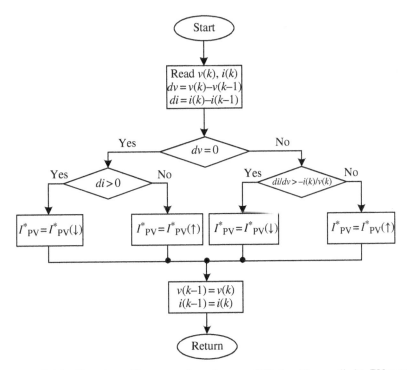

FIGURE 12.34   Flowchart of incremental conductance (IC) algorithm applied to PV current.

12.6.3   Beta Technique

The beta technique is a mathematical approximation to track the MPP by means of Equation 12.43, which is based on the PV module parameters, such as the electron charge (q), the diode ideality factor (a), the Boltzmann constant (k), the temperature over the PV surface (T_{pv}), and the number of PV cells in series (N_s).

$$\beta = \ln\left(\frac{I_{PV}}{V_{PV}}\right) - c.V_{PV} \, | c = \frac{q}{a.k.T_{pv}.N_s} \qquad (12.43)$$

The value of β at MPP is almost constant. Thus, under environmental changes, the β value can be digitally and continuously processed using the measured PV output quantities and inserted on a conventional closed-loop with constant reference (β^*), as shown in Figure 12.35.

12.6.4   Laboratory Project: Implementing the IC Technique

In this laboratory project, the IC technique algorithm is implemented in PSIM with fixed incremental step, as described in Figure 12.34. As shown in Figure 12.36, the standard PSIM solar module (physical model) has been used to represent six polycrystalline PV modules, Pluto Wde-240 [19] from Suntech, totaling 1440 W, 8.11 A_{SC}, 221.4 V_{OC}, and 177.6 V_{MPP}. The association of a DC–DC converter and its inner control loop has been modeled as a current-controlled source. The IC algorithm was written in custom C code and implemented into C block in PSIM platform. In Figure 12.37, the solar radiation (S) was steeply varied, allowing us to evaluate the MPPT.

Using the same electrical circuit of Figure 12.36, the student should implement the P&O and the beta MPPT techniques. At the end, compare the steady-state and dynamic behavior of these three MPPT techniques.

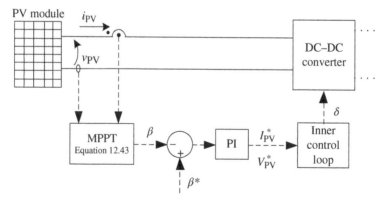

FIGURE 12.35   Beta technique with DC–DC converter in a PV power system.

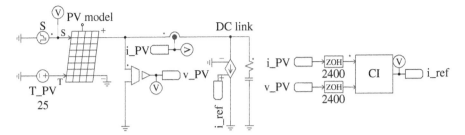

FIGURE 12.36   CI technique algorithm implemented in PSIM.

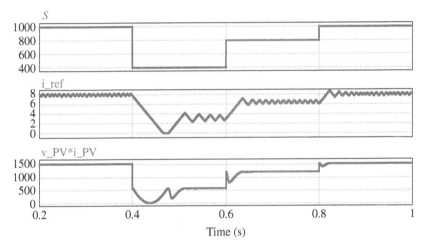

FIGURE 12.37   From top to bottom: solar radiation, PV current reference, and PV output power.

Customized C code implemented in C block for incremental conductance algorithm

```
// variables for IC
static double V, I, delta = 0.05, dI, dV;
static double I_=0, V_=0, Iref = 4;
I = in[0];                    //PV current input signal
V = in[1];                    //PV voltage input signal
dV = V - V_;                    //voltage difference
dI = I - I_;                  //current difference

// IC algorithm
if (dV == 0)
  { if (dI == 0) { }
    else
    {if (dI > 0) Iref = Iref - delta;
    else Iref = Iref + delta;
    }
  }
```

```
else
{ if (dI/dV == -I/V) { }
  else
  {if (dI/dV > -I/V) Iref = Iref - delta;
  else Iref = Iref + delta;
  }
}
I_ = I;                          //unit delayed current signal
V_ = V;                          //unit delayed voltage signal
if (Iref > 10) Iref =10;         //upper limiter
else if (Iref < 0) Iref = 0;     //lower limiter
out[0] = Iref;                   //output current reference signal
```

12.7 ISLANDING DETECTION

Islanding detection algorithms aim to sense the grid absence or its inappropriate operation, in order to isolate any distributed generation system from the mains, through a
proper circuit breaker. With this purpose, the islanding detection algorithm sends a
trigger signal to the circuit breaker and to the DG's control scheme, which should change
the DG operation state or shut it down. While the MPPT techniques are important for a
system efficiency, the islanding detection techniques are required for system protection
and user safety, especially during power system maintenance or failure.

Basically, the islanding detection techniques can be classified into three groups:
passive, active, and remote techniques. The passive techniques are based on the
measurement of local quantities. The active techniques commonly use the concept of
perturb and observe, which means that the DG system injects a disturbing signal into
the grid in order to sense its feedback response. If the feedback response is not as
expected, it means an islanding situation. In order to speed up the islanding detection, the remote techniques usually make use of communication unit, which allows
them to measure nonlocal quantities. All of these techniques show advantages and
drawbacks such as cost, accuracy, time response, and effectiveness.

Several standards have been formulated in order to guarantee the proper operation of loads and safety of users and to limit electrical disturbance propagation into
the grid. As an example, the IEEE 1547 standard for interconnection of distributed
resources with an electric power system [20], which is applied to DG with capacity
less than or equal to 30 kW (60 Hz) and establishes the conditions for the DG system
islanding and reconnection to the grid. If any RMS or fundamental frequency
values of each phase voltage is within a range given in Table 12.2, then DG must be
isolated with the corresponding maximum clearing time[2] indicated to the particular
range condition. The DG system must keep the islanded operation up to the instant
that all the reconnection constraints are attained, which means that the magnitude,

---

[2] Clearing time is the period from grid fault occurrence to actual turnoff of circuit breaker, including the
detection time, adjustable time, and intrinsic time delay of circuit breaker.

frequency, and phase deviations of voltage (difference between grid and DG voltages) are within the range given in Table 12.2, at least for 5 min (300 s).

12.7.1 Laboratory Project: Passive Islanding Detection Based on IEEE Std. 1547

As a laboratory project, let us digitally implement a passive islanding detection algorithm that senses the voltage at the DG's connection point and compares it with the IEEE Std. 1547. If the voltage parameters are within acceptable limits, the DG system continues to operate connected to the grid, whereas if the voltage parameters do not attain the requirements, the islanding detection algorithm must inform the DG's control and, concomitantly, open the circuit breaker responsible to isolate the system from the mains, as shown in Figure 12.38a. Thus, the islanding detection algorithm is modeled as a subsystem, as shown in Figure 12.38b, responsible to define the islanding and grid-connected operation by means of a trigger signal sent to the circuit breaker and to the DG's control scheme. The grid and DG voltages are measured and processed through a PLL algorithm, as described in Section 12.5. From there, the quantities are compared with the ranges listed in Table 12.2 and the trigger

TABLE 12.2 Standard of Operation for System Islanding and Reconnecting to the Grid, GD ≤ 30 kW

Islanding (Maximum Clearing Time)		Reconnecting (Minimum Steady State Time)		
Voltage Level (%)	Frequency (Hz)	Δ Voltage (%)	Δ Frequency (Hz)	Δ Phase (°)
$V < 50$ (0.16 s) $50 \le V < 88$ (2.00 s)	>60.5 (0.16 s)			
$110 < V \le 120$ (1.00 s) $V > 120$ (0.16 s)	<59.3 (0.16 s)	<10 (300 s)	<0.3 (300 s)	<20 (300 s)

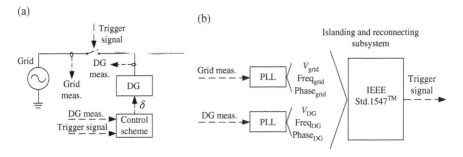

FIGURE 12.38 Islanding and reconnecting subsystem based on the IEEE Std. 1547. (a) DG measurement with trigger signal for commanding and islanding switch, (b) PLL measurements of grid voltage, frequency and phase for commanding islanding and reconnection.

signal is set. By definition, the trigger signal equal to unitary value means DG connected to the grid, while null trigger signal means islanded operation.

The system illustrated in Figure 12.38 was implemented in PSIM as shown in Figure 12.39, where the DG system is modeled as a voltage source. The RMS and PLL algorithms were previously described in Sections 12.2.3 and 12.5, respectively. The islanding detection block, based on IEEE Std. 1547, was written in custom C code. The full implementation can be found in the simulation files at the webpage maintained by the authors. Note that the requested maximum clearing time and minimum steady-state time of Table 12.2 were reduced for the sake of better visualization. Thus, a grid voltage reduction of 55% was applied at 0.3 s and restored to normal operation at 0.4 s. From the generated result shown in Figures 12.40 and 12.41, we observe the proper islanding and reconnection procedures complying with the defined adjustable time delay. Time 1 corresponds to the time interval required for the trigger signal be zero (disconnect), while Time 4 is the time interval needed for the trigger signal returns to one (re-connect).

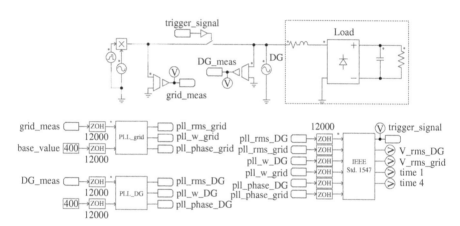

FIGURE 12.39 Islanding detection algorithm implemented in PSIM.

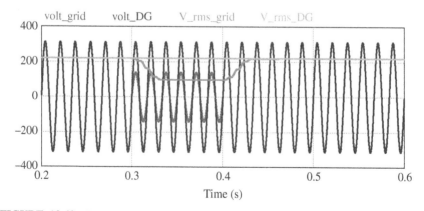

FIGURE 12.40 Instantaneous and RMS voltages of grid and DG sides during voltage reduction.

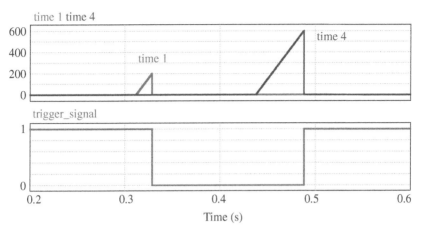

FIGURE 12.41 Results using the passive islanding detection: adjustable time delays and output trigger signals.

12.8 SUGGESTED PROBLEMS

1. Implement and evaluate a custom C code for calculating the derivative function represented in Equation 12.2.

2. Implement in PSIM and compare the three integrator methods described in Section 12.2.2: backward Euler, forward Euler, and trapezoidal.

3. Write the customized C code for implementing the PI controller with anti-windup, as shown in Figure 12.42. Consider that the L_p is the limit for the proportional part, and L_I is the dynamic integrator limiter calculated as in Equation 12.44. K_p and K_I are, respectively, the proportional and integral gains.

$$|L_I(k)| = L_p - |K_p \cdot \varepsilon_x(k)|. \tag{12.44}$$

4. Implement in PSIM and compare two MAFs, one with time response of one fundamental cycle and a second one with five cycles, as in Section 12.2.3.

5. The reactive current is defined as

$$i_r = \frac{W}{\hat{V}^2} \cdot \hat{v}, | W = \frac{1}{T} \cdot \int_{t-T}^{t} \hat{v} \cdot i d\tau, \tag{12.45}$$

such that $\hat{v}$ and $\hat{V}$ are, respectively, the instantaneous and RMS unbiased time integral of voltage, and W is the average reactive power. Thus, similar to Section 12.2.4, calculate the reactive current performing the load variation (R, L, C) analysis.

6. Design, simulate, and analyze an IIR-tuned filter, assuming a center frequency of 50 Hz, with sampling frequency of 10 kHz and stop band of 4 Hz, similar to that depicted in Figure 12.11.

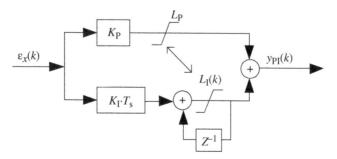

FIGURE 12.42   Block diagram of a PI controller with anti-windup limitation.

7. Using the procedure suggested in Section 12.4.2, repeat the laboratory project from Section 12.4.3 and calculate the symmetrical components through the modified FIR filter.

8. Repeat the laboratory project from Section 12.5.3 applying a step variation on the magnitude of the input signal, instead of a jump phase.

9. Implemented the THD algorithm of Section 12.3.3 using a FWD based on PLL, as in Section 12.5.4, instead of IIR filter.

10. As discussed in Section 12.4, sequence components can be calculated after the application of fundamental components identification, either using IIR or FIR filters. However, if fundamental frequency deviations are expected, the application of adaptive filters would be required in order to ensure effective filtering. Other possibility is using the positive-sequence detector (PSD) presented in [16], which is based on a PLL to synchronize to eventual frequency variation.

Thus, the PLL is responsible for identifying the phase angle of phase "*a*" and the following unitary sinusoids may be generated, so that each one is in-phase to their respective input phase voltages:

$$\boldsymbol{y} = \begin{bmatrix} y_a \\ y_b \\ y_c \end{bmatrix} = \begin{bmatrix} \cos(\theta) \\ \cos(\theta - 120°) \\ \cos(\theta + 120°) \end{bmatrix}. \tag{12.46}$$

Therefore, the scalar product of such sinusoids and the input signals would result as follows:

$$\boldsymbol{x} \cdot \boldsymbol{y} = x_a y_a + x_b y_b + x_c y_c = \bar{c} + \tilde{c}. \tag{12.47}$$

Here, the constant value ($\bar{c}$) is proportional to the amplitude of the positive sequence and may be extracted by means of a MAF.

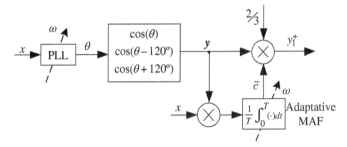

FIGURE 12.43   Fundamental positive-sequence detector.

$$
y_1^+ = \begin{bmatrix} y_{a_1}^+ \\ y_{b_1}^+ \\ y_{c_1}^+ \end{bmatrix} = \overline{c} \cdot \frac{2}{3} \cdot \begin{bmatrix} y_a \\ y_b \\ y_c \end{bmatrix} = \overline{c} \cdot \frac{2}{3} \cdot \begin{bmatrix} \cos(\theta) \\ \cos(\theta - 120°) \\ \cos(\theta + 120°) \end{bmatrix}. \tag{12.48}
$$

Figure 12.43 depicts the process of positive-sequence detection by means of the PLL. It is interesting to note that the algorithm is immune to frequency deviations, as well as to waveform distortions. You should adapt such algorithm to calculate the negative-sequence components and then implement such technique in PSIM or Matlab software, either using blocks or C code.

REFERENCES

[1] DESTRO, R., MATAKAS, L., KOMATSU, W. and AMA, N.R.N., "Implementation aspects of adaptive window moving average filter applied to PLLs—comparative study", in Brazilian Power Electronics Conference (COBEP), Gramado, IEEE, pp. 730–736, 2013.

[2] TENTI, P., PAREDES, H.K.M. and MATTAVELLI, P., "Conservative power theory, a framework to approach control and accountability issues in smart microgrids", IEEE Transactions on Power Electronics, vol. 26, no. 3, pp. 664–673, 2011.

[3] CAVALLINI, A. and MONTANARI, G.C., "Compensation strategies for shunt active-filter control", IEEE Transactions on Power Electronics, vol. 9, no. 6, pp. 587–593, 1994.

[4] AREDES, M,, HAFNER, J. and HEUMANN, K., "Three-phase four-wire shunt active filter control strategies", IEEE Transactions on Power Electronics, vol. 12, no. 2, pp. 311–318, 1997.

[5] MARAFÃO, F.P., BRANDÃO, D.I., GONÇALVES, F.A.S. and PAREDES, H.K.M., "Decoupled reference generator for shunt active filters using the conservative power theory", Journal of Control, Automation and Electrical Systems, vol. 24, no. 4, pp. 522–534, 2013.

[6] OPPENHEIM, A.V., SCHAFER, R.W. and BUCK, J.R., Discrete-Time Signals Processing, 2nd edition, Prentice Hall, Englewood Cliffs, 1999.

[7] NEWMAN, M.J. and HOLMES, D.G., "Delta operator digital filters for high performance inverter applications", IEEE Transactions on Power Electronics, vol. 18, no. 1, pp. 447–454, 2003.

[8] MARAFÃO, F.P., DECKMANN, S.M. and LOPES, A., "Robust delta operator-based discrete systems for fixed-point DSP implementations", in Nineteenth Annual IEEE Applied Power Electronics Conference and Exposition, 2004. APEC '04, IEEE, pp. 1764–1770, 2004.

[9] MATTAVELLI, P. and MARAFÃO, F.P., "Repetitive-based control for selective harmonic compensation in active power filters", IEEE Transactions on Industrial Electronics, vol. 51, no. 5, pp. 1018–1024, 2004.

[10] FORTESCUE, C.L., "Method of symmetrical co-ordinates applied to the solution of polyphase networks", Transactions of the American Institute of Electrical Engineers, vol. 37, no. 2, pp. 1027–1140, 1918.

[11] KAURA, V. and BLASKO, V., "Operation of a phase locked loop system under distorted utility conditions", IEEE Transactions on Industry Applications, vol. 33, no. 1, pp. 58–63, 1997.

[12] DA SILVA, S.A.O, GARCIA, P.F.D, CORTIZO, P.C. and SEIXAS, P.F., "A three-phase line-interactive UPS system implementation with series-parallel active power-line conditioning capabilities", IEEE Transactions on Industry Applications, vol. 38, no. 6, pp. 1581–1590, 2002.

[13] GOLESTAN, S., RAMEZANI, M., GUERRERO, J.M., FREIJEDO, F.D. and MONFARED, M., "Moving average filter based phase-locked loops: performance analysis and design guidelines", IEEE Transactions on Power Electronics, vol. 29, no. 6, pp. 2750–2763, 2014.

[14] BRANDÃO, D.I., MARAFÃO, F.P., SIMÕES, M.G. and POMILIO, J.A., "Considerations on the modeling and control scheme of grid connected inverter with voltage support capability", in Brazilian Power Electronics Conference, Gramado, 2013.

[15] OGATA, K., Modern Control Engineering, 5th edition, Prentice Hall, Boston, 2010.

[16] PÁDUA, M.S., DECKMANN, S.M. and MARAFÃO, F.P., "Frequency-adjustable positive sequence detector for power conditioning applications", in IEEE Power Electronics Specialists Conference, IEEE, pp. 1928–1934, 2005.

[17] DE BRITO, M.A.G., GALOTTO, L., SAMPAIO, L.P., DE MELO, G.A. and CANESIN, C.A., "Evaluation of the main MPPT techniques for photovoltaic applications", IEEE Transactions on Industrial Electronics, vol. 60, no. 3, pp. 1156–1167, 2013.

[18] HSIEH, G.C., CHEN, H.L., CHEN, Y., TSAI, C.M. and SHYU, S.S., "Variable frequency controlled incremental conductance derived MPPT photovoltaic stand-along DC bus system", in IEEE Applied Power Electronics Conference and Exposition, IEEE, pp. 1849–1854, 2008.

[19] Suntech, HiPerforma module PLUTO 240-Wde Polycrystalline solar module datasheet. http://www.solarchoice.net.au/blog/wp-content/uploads/PLUTO240-245-Hiperforma-Suntech-Solar-Panels.pdf (accessed May 10, 2016).

[20] IEEE "IEEE standard for interconnecting distributed resources with electric power systems", IEEE Std 1547, 2003.

INDEX

absolute value amplifier, 134–135
acausal system, 63, 64
active power, 21, 29, 35–37, 57, 182–186, 190, 191, 193, 196, 197, 227, 229, 241, 242, 257, 259–261, 267, 270, 279, 280, 286, 287, 310, 317
A/D converter, 265, 267, 269, 270
air permeability, 123
analog computer, 2, 142, 143, 145
analog- or digital-based modeling, 4
analytical solutions, 67
analyzer, 230
angular speed, 107, 109, 111, 114, 152, 158
armature current, 111, 158
armature voltage, 111
automatic code generation, 255, 264–277

balance of charges, 16
balance of fluxes, 16
Ballard mark, 213, 214
bang–bang, 179
beta technique, 310, 312

block diagram, 3, 4, 6–8, 17–19, 21, 23, 24, 37, 43–58, 61, 63, 72, 85, 86, 93, 107, 118, 144, 145, 148, 163, 185, 235, 251, 301, 302, 306, 318
block diagram-oriented simulator, 18
Bode, 86, 94, 97–99, 112, 145, 231, 288
boost converter, 15, 16, 88–91, 93, 97–106, 154, 190, 192, 193, 205, 208, 257, 258, 260, 262, 273, 275
buck converter, 89, 154, 207, 208

capacitive sensors, 121–123
cascaded loop, 179
CASPOC, 17, 18
circuit-oriented simulation, 61–81
circuit-oriented simulator, 18, 21
circuit simulator, 3, 8, 61–63, 72, 140
Clarke transformation, 188
clipped detector, 126
closed-loop control, 12, 86–88, 107, 145, 178
closed-loop transfer function, 12, 302
CMMR (common-mode rejection ratio), 128

Modeling Power Electronics and Interfacing Energy Conversion Systems, First Edition.
M. Godoy Simões and Felix A. Farret.
© 2017 John Wiley & Sons, Inc. Published 2017 by John Wiley & Sons, Inc.

common coupling, 72, 179, 224, 227
component-level, 62
computer-based simulation, 2, 143
constant current control, 181–182
constant frequency, 89
constant voltage control, 131, 140, 178, 181, 185, 190, 257, 259
current source loops, 6
cut-set, 6
cylindrical inductor, 124

D/A converter, 265, 267, 269, 270
DAE (differential algebraic equation), 2, 3, 17, 18, 63
data acquisition, 118–119, 123, 125–139, 238
data storage, 49
DCT filter, 297–298
derivative, 2, 5, 6, 14, 18, 47, 48, 62, 88, 111, 138, 280–286, 289, 310, 317
DFIG (doubly fed induction generator), 154, 173, 208, 209
DFT (discrete Fourier transform), 3, 235–239, 242–246
DG (distributed generation), 3, 4, 177–179, 184, 190, 191, 227, 229, 242, 252, 280, 300, 301, 314–316
differential amplifier, 117, 125, 127, 128
differential output, 128
differentiation, 2, 125
digital filter, 113, 279, 280, 288, 289
digital processing, 2, 279–319
discrete control, 107–111
discrete PI control, 88
divider, 30, 97, 119, 134, 141, 142, 256, 270
DOD (depth of discharge), 218
dq0–abc, 181
d–q axis, 179, 182, 191
DSP, 62, 88, 130, 191, 255–277
DYMOLA, 17
dynamic system, 9

ECAP, 17
electrical torque, 111
element library, 256, 257
energy conversion, 3, 13, 17, 18, 43, 177, 250
energy storage elements, 6
energy systems, 3, 4, 12–18, 44, 116–145, 158, 175, 197, 207
equivalent circuit, 13, 18, 37–39, 90, 114, 136, 148, 154, 158, 162, 207, 211, 215

equivalent Laplace, 45
equivalent mechanical, 207
error amplifier, 88, 93–96, 140, 143, 145
Euler's method, 2

fault detection, 228
feedback control, 85, 86, 181, 184
feedback function, 12
FET, 37, 39, 131
FFT (fast Fourier transform), 3, 235, 237–239, 257, 287, 291
finite element analysis, 256
FIR filter, 288, 290–292, 297, 298, 318
first-order differential equations, 4
first-order equations, 6, 46
flyback circuit, 15
Fortescue, 293–296
Fourier series, 231–237, 290
Fourier transform, 3, 234, 235, 237–239, 287, 291
frequency regulation, 177
fuel cell, 43, 89, 178, 203, 211–215, 224
full-fledged integration, 61
full-wave rectifier, 134
fundamental component, 169, 243, 287–293, 296, 298, 318

Google Group, power-electronics-interfacing-energy-conversion-systems@googlegroups.com, xiv, 250
grid-connected inverter, 159, 177–199
grid synchronization, 182, 188, 199, 279

half-wave rectifier, 13, 14, 63, 67–71, 134
high-gain amplifier, 136, 143
hybrid mathematical model, 17
hydropower, 43, 178
hysteresis-band, 131, 179

ideal switches, 13
IEEE, 179, 184–187, 199, 227, 228, 314–316
IGBT, 13
IIR filter, 288–290, 292, 293, 296–299, 318
induction machine, 38, 148–152, 154–156, 175, 205, 207
inductive sensor, 123–124
insertion capacitive sensor, 122
instrumentation, 117–145, 230
instrumentation amplifier, 127, 128
integrated power plant, 219–224

integration, 2, 8, 18, 61, 65, 66, 70, 125, 129, 139, 301
integrator, 1, 7, 8, 188, 260, 281, 282, 302, 306, 317
inverting input, 129, 131, 138
islanding, 178, 184, 198, 227, 279, 280, 314–317

KCL equations, 7, 30
K method, 95–96
KVL loop, 7

laboratory project, 4, 29–37, 45–52, 72–78, 107–111, 138, 140, 158–169, 190–197, 242–246, 286, 291, 298, 303, 306, 312, 315, 318
Laplace operator, 28, 47, 88
Laplace transform, 8–10, 45–52, 86, 92, 112, 235
LDR (light-dependent resistor), 121
lead acid battery, 184, 215–219, 223
level detector, 126, 131
level measurement, 123
library browser, 257
linear system, 3, 27, 28, 33, 54
load torque, 109, 111
lossless, 89, 90

MAF (moving average filter), 279, 280, 282–286, 290, 301–305, 317–319
magnetic core, 45, 58, 123, 124
magnetic reluctance, 123, 124
magnetic sensor, 124
Maple, 48, 65, 231
Mathematica, 48, 65, 231
mathematical-oriented model, 2
MATLAB, 3, 6, 8, 10–12, 21, 29–31, 35, 37–39, 52, 56–58, 61–63, 65–68, 71–78, 97, 98, 107–111, 114, 116, 118, 142, 163, 235, 237, 242–246, 250–252
MATLAB/Simulink, 8, 79, 142, 148, 163, 224, 256
matrix, 3–7, 10, 17, 18, 27, 28, 30–33, 46, 48, 58, 66, 295
matrix equations, 7, 28, 48
matrix form, 5, 6, 33, 46
mechanical systems, 85, 113, 114
mesh analysis, 6, 27, 29, 32, 34, 35
$m \times n$ matrix, 3

MODELICA, 17
modified nodal analysis, 18
modular approach, 63
motor driver, 156–158
MPPT (maximum power point tracking), 190, 191, 197, 279, 280, 307–310, 312, 314
multiplier, 217, 255

network topology, 27
Newton–Raphson method, 18
nodal analysis, 18, 27–40, 48
non inverting input, 121, 132, 138, 141
nonlinearity, 45, 57, 58, 86, 120, 121, 128–130
nonlinearity compensation, 129
NTC, 120
nth-order differential equation, 4

object-oriented programming, 4
ODE (ordinary differential equation), 1–3, 6, 17, 18, 43, 63
one-diode model, 204
operational amplifiers, 23, 88, 103, 117, 118, 125, 128, 129, 131, 132, 134, 138, 140–143, 145, 259, 3879
optimal control, 64
optotransistor, 129
output-isolating amplifier, 130

PCC (point of common coupling), 179, 181–185, 190, 197, 224, 227, 251, 252
peak detector, 134
PEM (proton exchange membrane), 211
permanent magnet synchronous generator, 158, 208
phase-locked loop, 118, 178, 188–189
phase shift, 35, 57, 86, 125, 131–134, 149, 178, 179, 188, 231, 248, 249, 252
phasor analysis, 28, 29, 46, 52
phasor-based, 35, 52–54
photocoupler, 128–130
photovoltaic, 38, 39, 43, 89, 115, 178, 192, 203–205, 224, 307
physics-based model, 4, 44
PI (proportional integral), 23, 88, 107, 109, 143–145, 179, 181, 189–191, 301, 302, 317, 318
PI control, 23, 88, 107, 109, 143, 145, 179, 181, 189–191, 260, 301, 302, 317, 318

PID (proportional integral derivative), 88, 280
PIL (processor-in-the-loop), 255, 272–277
PLECS, 17
PLL (phase-locked loop), 178, 179, 181, 182, 187–189, 199, 288, 300–307, 315, 316, 318, 319
PMSG (permanent magnet synchronous generator), 158–160, 208–210, 222
poles and zeros, 96
potentiometer, 120, 121
power quality, 3, 13, 21, 44, 118, 175, 178, 179, 181, 224, 227–252, 279, 287, 291, 298
power systems toolbox, 3, 17, 18, 21, 61, 62, 180, 181, 224
p–q, 179, 182, 183
precision diode, 134
precision rectifier, 134, 135
P+resonant, 179, 187–189
probe and meter, 256
problem-based learning, 4
problem-based learning strategies, 4
processor-in-the-loop, 255, 256, 272–277
project-based learning, 4
proportional control, 19, 304
PSIM, 3, 17, 18, 21, 29, 37, 61–63, 67, 68, 70–80, 99, 103, 113–116, 118, 126, 129, 137, 138, 140–143, 148, 150–153, 158, 169, 190, 192, 193, 195, 197, 199, 204, 207, 211, 214, 219, 224, 243, 244, 246, 250, 251, 252, 255–277, 281, 284–286, 292, 298, 303, 306, 312, 313, 316, 317, 319
PTC, 120
push–pull amplifier, 79, 137
P–V control, 183–184
PWM (pulse width modulation), 18, 19, 21, 23, 89, 91, 93, 97, 98, 103, 131, 137–140, 143–145, 154, 169, 178, 181, 182, 188, 191, 256, 257, 265, 268, 269, 271

QR factorization, 3

reactive power, 29, 36, 57, 182–185, 190, 191, 193, 197, 242, 257, 259–261, 267, 270, 317
real-time control, 18, 62
real-time information, 178
renewable energy, 3, 4, 44, 89, 118, 178, 191, 219, 307
residual magnetism, 156, 207

Riccati equation, 64
rotating machine generators, 43
rotational-based capacitive sensor, 122
Runge–Kutta method, 2

Schmitt trigger, 131, 138
SCI block, 265, 271–272
SCR (short circuit ratio), 242
s-domain, 188, 256, 259
semiconductor switch, 89, 255, 256
sensor, 84, 85, 117–125, 145, 178, 255, 256, 267
sequence component, 293–300, 318
series resonant, 56
shaft speed, 109, 111
signal processing, 18, 28, 52, 64, 117, 118, 188, 235, 279, 280
SIMPLORER, 17
Simscape blocks, 118
simulation block diagram, 7, 8
simulation plan, 72
simulator, 2, 3, 6, 8, 16, 18, 21, 61–63, 72, 118, 140
SISO system, 10
small signal, 90–94
smart meter, 178
SOC (state of charge), 115, 215, 216, 218, 219, 224
SPICE (simulation program with integrated circuit emphasis), 17, 18, 62, 140
SPWM (sinusoidal pulse width modulation), 114, 115, 138–140, 158–169, 172, 188, 191
SRF (synchronous reference frame), 188
stand-alone inverter, 181, 190–197
state equation, 4, 8, 10, 11, 17, 56, 91
state space, 4–7, 10–12, 17, 28, 46–49, 76, 158, 219
state space based modeling, 4–6
state-space equation, 4, 7, 10, 11, 47
state-space formulation, 4, 6, 7, 46–48, 219
state space to transfer function conversion, 10–12
state variable, 2, 5–8, 13, 14, 18, 46–48, 90
stationary frame, 181, 187–188, 301
steady-state analysis, 52–54
superposition theorem, 29–37
switches, 13, 15–17, 44, 99, 137, 186, 197, 255, 256
synchronous speed, 148, 205

temperature coefficient, 119, 120, 223
THD (total harmonic distortion), 230, 241, 243, 247, 252, 279, 287, 291–293, 318
thermistor, 120
Thévenin, 6, 37, 52, 54–55
time domain, 8, 47, 52, 55, 56, 64, 86, 88, 234, 235, 238, 240, 287
time varying, 17, 73, 216
transducer, 118–124, 230
transfer function, 9–12, 61–63, 88, 90–98, 112–114, 191, 219, 256, 259, 281, 283, 302
transfer function conversion, 10–12
transformation star delta, 136
trap filter, 56
trial and error, 83, 88

triangular waveform, 137–139, 145
trigger ADC, 269

virtual ground, 24, 132
voltage crossover, 79, 126
voltage quality, 228, 229
voltage regulation, 177, 190, 191, 224
voltage source loops, 6
voltage zero crossing, 126, 137

wind power, 207–210, 220, 224
workspace, 49, 51, 52, 78, 107, 109, 251

z-domain, 256, 260, 289
zener diode, 126